D0806063

Engineering Quality by Design

STATISTICS: Textbooks and Monographs

Vol. 1: The Generalized Jackknife Statistic, *H. L. Gray and W. R. Schucany*
Vol. 2: Multivariate Analysis, *Anant M. Kshirsagar*
Vol. 3: Statistics and Society, *Walter T. Federer*
Vol. 4: Multivariate Analysis: A Selected and Abstracted Bibliography, 1957-1972, *Kocherlakota Subrahmaniam and Kathleen Subrahmaniam* (out of print)
Vol. 5: Design of Experiments: A Realistic Approach, *Virgil L. Anderson and Robert A. McLean*
Vol. 6: Statistical and Mathematical Aspects of Pollution Problems, *John W. Pratt*
Vol. 7: Introduction to Probability and Statistics (in two parts), Part I: Probability; Part II: Statistics, *Narayan C. Giri*
Vol. 8: Statistical Theory of the Analysis of Experimental Designs, *J. Ogawa*
Vol. 9: Statistical Techniques in Simulation (in two parts), *Jack P. C. Kleijnen*
Vol. 10: Data Quality Control and Editing, *Joseph I. Naus* (out of print)
Vol. 11: Cost of Living Index Numbers: Practice, Precision, and Theory, *Kali S. Banerjee*
Vol. 12: Weighing Designs: For Chemistry, Medicine, Economics, Operations Research, Statistics, *Kali S. Banerjee*
Vol. 13: The Search for Oil: Some Statistical Methods and Techniques, *edited by D. B. Owen*
Vol. 14: Sample Size Choice: Charts for Experiments with Linear Models, *Robert E. Odeh and Martin Fox*
Vol. 15: Statistical Methods for Engineers and Scientists, *Robert M. Bethea, Benjamin S. Duran, and Thomas L. Boullion*
Vol. 16: Statistical Quality Control Methods, *Irving W. Burr*
Vol. 17: On the History of Statistics and Probability, *edited by D. B. Owen*
Vol. 18: Econometrics, *Peter Schmidt*
Vol. 19: Sufficient Statistics: Selected Contributions, *Vasant S. Huzurbazar (edited by Anant M. Kshirsagar)*
Vol. 20: Handbook of Statistical Distributions, *Jagdish K. Patel, C. H. Kapadia, and D. B. Owen*
Vol. 21: Case Studies in Sample Design, *A. C. Rosander*
Vol. 22: Pocket Book of Statistical Tables, *compiled by R. E. Odeh, D. B. Owen, Z. W. Birnbaum, and L. Fisher*
Vol. 23: The Information in Contingency Tables, *D. V. Gokhale and Solomon Kullback*

Vol. 24: Statistical Analysis of Reliability and Life-Testing Models: Theory and Methods, *Lee J. Bain*

Vol. 25: Elementary Statistical Quality Control, *Irving W. Burr*

Vol. 26: An Introduction to Probability and Statistics Using BASIC, *Richard A. Groeneveld*

Vol. 27: Basic Applied Statistics, *B. L. Raktoe and J. J. Hubert*

Vol. 28: A Primer in Probability, *Kathleen Subrahmaniam*

Vol. 29: Random Processes: A First Look, *R. Syski*

Vol. 30: Regression Methods: A Tool for Data Analysis, *Rudolf J. Freund and Paul D. Minton*

Vol. 31: Randomization Tests, *Eugene S. Edgington*

Vol. 32: Tables for Normal Tolerance Limits, Sampling Plans, and Screening, *Robert E. Odeh and D. B. Owen*

Vol. 33: Statistical Computing, *William J. Kennedy, Jr. and James E. Gentle*

Vol. 34: Regression Analysis and Its Application: A Data-Oriented Approach, *Richard F. Gunst and Robert L. Mason*

Vol. 35: Scientific Strategies to Save Your Life, *I. D. J. Bross*

Vol. 36: Statistics in the Pharmaceutical Industry, *edited by C. Ralph Buncher and Jia-Yeong Tsay*

Vol. 37: Sampling from a Finite Population, *J. Hajek*

Vol. 38: Statistical Modeling Techniques, *S. S. Shapiro*

Vol. 39: Statistical Theory and Inference in Research, *T. A. Bancroft and C.-P. Han*

Vol. 40: Handbook of the Normal Distribution, *Jagdish K. Patel and Campbell B. Read*

Vol. 41: Recent Advances in Regression Methods, *Hrishikesh D. Vinod and Aman Ullah*

Vol. 42: Acceptance Sampling in Quality Control, *Edward G. Schilling*

Vol. 43: The Randomized Clinical Trial and Therapeutic Decisions, *edited by Niels Tygstrup, John M. Lachin, and Erik Juhl*

Vol. 44: Regression Analysis of Survival Data in Cancer Chemotherapy, *Walter H. Carter, Jr., Galen L. Wampler, and Donald M. Stablein*

Vol. 45: A Course in Linear Models, *Anant M. Kshirsagar*

Vol. 46: Clinical Trials: Issues and Approaches, *edited by Stanley H. Shapiro and Thomas H. Louis*

Vol. 47: Statistical Analysis of DNA Sequence Data, *edited by B. S. Weir*

Vol. 48: Nonlinear Regression Modeling: A Unified Practical Approach, *David A. Ratkowsky*

Vol. 49: Attribute Sampling Plans, Tables of Tests and Confidence Limits for Proportions, *Robert E. Odeh and D. B. Owen*

Vol. 50: Experimental Design, Statistical Models, and Genetic Statistics, *edited by Klaus Hinkelmann*

Vol. 51: Statistical Methods for Cancer Studies, *edited by Richard G. Cornell*

Vol. 52: Practical Statistical Sampling for Auditors, *Arthur J. Wilburn*

Vol. 53: Statistical Signal Processing, *edited by Edward J. Wegman and James G. Smith*

Vol. 54: Self-Organizing Methods in Modeling: GMDH Type Algorithms, *edited by Stanley J. Farlow*

Vol. 55: Applied Factorial and Fractional Designs, *Robert A. McLean and Virgil L. Anderson*

Vol. 56: Design of Experiments: Ranking and Selection, *edited by Thomas J. Santner and Ajit C. Tamhane*

Vol. 57: Statistical Methods for Engineers and Scientists. Second Edition, Revised and Expanded, *Robert M. Bethea, Benjamin S. Duran, and Thomas L. Boullion*

Vol. 58: Ensemble Modeling: Inference from Small-Scale Properties to Large-Scale Systems, *Alan E. Gelfand and Crayton C. Walker*

Vol. 59: Computer Modeling for Business and Industry, *Bruce L. Bowerman and Richard T. O'Connell*

Vol. 60: Bayesian Analysis of Linear Models, *Lyle D. Broemeling*

Vol. 61: Methodological Issues for Health Care Surveys, *Brenda Cox and Steven Cohen*

Vol. 62: Applied Regression Analysis and Experimental Design, *Richard J. Brook and Gregory C. Arnold*

Vol. 63: Statpal: A Statistical Package for Microcomputers – PC-DOS Version for the IBM PC and Compatibles, *Bruce J. Chalmer and David G. Whitmore*

Vol. 64: Statpal: A Statistical Package for Microcomputers – Apple Version for the II, II+, and IIe, *David G. Whitmore and Bruce J. Chalmer*

Vol. 65: Nonparametric Statistical Inference, Second Edition, Revised and Expanded, *Jean Dickinson Gibbons*

Vol. 66: Design and Analysis of Experiments, *Roger G. Petersen*

Vol. 67: Statistical Methods for Pharmaceutical Research Planning, *Sten W. Bergman and John C. Gittins*

Vol. 68: Goodness-of-Fit Techniques, *edited by Ralph B. D'Agostino and Michael A. Stephens*

Vol. 69: Statistical Methods in Discrimination Litigation, *edited by D. H. Kaye and Mikel Aickin*

Vol. 70: Truncated and Censored Samples from Normal Populations, *Helmut Schneider*

Vol. 71: Robust Inference, *M. L. Tiku, W. Y. Tan, and N. Balakrishnan*

Vol. 72: Statistical Image Processing and Graphics, *edited by Edward J. Wegman and Douglas J. DePriest*

Vol. 73: Assignment Methods in Combinatorial Data Analysis, *Lawrence J. Hubert*

Vol. 74: Econometrics and Structural Change, *Lyle D. Broemeling and Hiroki Tsurumi*

Vol. 75: Multivariate Interpretation of Clinical Laboratory Data, *Adelin Albert and Eugene K. Harris*

Vol. 76: Statistical Tools for Simulation Practitioners, *Jack P. C. Kleijnen*

Vol. 77: Randomization Tests, Second Edition, *Eugene S. Edgington*

Vol. 78: A Folio of Distributions: A Collection of Theoretical Quantile-Quantile Plots, *Edward B. Fowlkes*

Vol. 79: Applied Categorical Data Analysis, *Daniel H. Freeman, Jr.*

Vol. 80: Seemingly Unrelated Regression Equations Models : Estimation and Inference, *Virendra K. Srivastava and David E. A. Giles*

Vol. 81: Response Surfaces: Designs and Analyses, *Andre I. Khuri and John A. Cornell*

Vol. 82: Nonlinear Parameter Estimation: An Integrated System in BASIC, *John C. Nash and Mary Walker-Smith*

Vol. 83: Cancer Modeling, *edited by James R. Thompson and Barry W. Brown*

Vol. 84: Mixture Models: Inference and Applications to Clustering, *Geoffrey J. McLachlan and Kaye E. Basford*

Vol. 85: Randomized Response: Theory and Techniques, *Arijit Chaudhuri and Rahul Mukerjee*

Vol. 86: Biopharmaceutical Statistics for Drug Development, *edited by Karl E. Peace*

Vol. 87: Parts per Million Values for Estimating Quality Levels, *Robert E. Odeh and D. B. Owen*

Vol. 88: Lognormal Distributions: Theory and Applications, *edited by Edwin L. Crow and Kunio Shimizu*

Vol. 89: Properties of Estimators for the Gamma Distribution, *K. O. Bowman and L. R. Shenton*

Vol. 90: Spline Smoothing and Nonparametric Regression, *Randall L. Eubank*

Vol. 91: Linear Least Squares Computations, *R. W. Farebrother*

Vol. 92: Exploring Statistics, *Damaraju Raghavarao*

Vol. 93: Applied Time Series Analysis for Business and Economic Forecasting, *Sufi M. Nazem*

Vol. 94: Bayesian Analysis of Time Series and Dynamic Models, *edited by James C. Spall*

Vol. 95: The Inverse Gaussian Distribution: Theory, Methodology, and Applications, *Raj S. Chhikara and J. Leroy Folks*

Vol. 96: Parameter Estimation in Reliability and Life Span Models, *A. Clifford Cohen and Betty Jones Whitten*

Vol. 97: Pooled Cross-Sectional and Time Series Data Analysis, *Terry E. Dielman*

Vol. 98: Random Processes: A First Look, Second Edition, Revised and Expanded, *R. Syski*

Vol. 99: Generalized Poisson Distributions: Properties and Applications, *P.C. Consul*

Vol. 100: Nonlinear L_p-Norm Estimation, *René Gonin and Arthur H. Money*

Vol. 101: Model Discrimination for Nonlinear Regression Models, *Dale S. Borowiak*

Vol. 102: Applied Regression Analysis in Econometrics, *Howard E. Doran*

Vol. 103: Continued Fractions in Statistical Applications, *K.O. Bowman and L.R. Shenton*

Vol. 104: Statistical Methodology in the Pharmaceutical Sciences, *Donald A. Berry*

Vol. 105: Experimental Design in Biotechnology, *Perry D. Haaland*

Vol. 106: Statistical Issues in Drug Research and Development, *edited by Karl E. Peace*

Vol. 107: Handbook of Nonlinear Regression Models, *David A. Ratkowsky*

Vol. 108: Robust Regression: Analysis and Applications, *edited by Kenneth D. Lawrence and Jeffrey L. Arthur*

Vol. 109: Statistical Design and Analysis of Industrial Experiments, *edited by Subir Ghosh*

Vol. 110: *U*-Statistics: Theory and Practice, *A. J. Lee*

Vol. 111: A Primer in Probability, Second Edition, Revised and Expanded, *Kathleen Subrahmaniam*

Vol. 112: Data Quality Control: Theory and Pragmatics, *edited by Gunar E. Liepins and V. R. R. Uppuluri*

Vol. 113: Engineering Quality by Design: Interpreting the Taguchi Approach, *Thomas B. Barker*

Vol. 114: Survivorship Analysis for Clinical Studies, *Eugene K. Harris and Adelin Albert*

ADDITIONAL VOLUMES IN PREPARATION

Engineering Quality by Design
Interpreting the Taguchi Approach

Thomas B. Barker

Graduate Statistics Department
Center for Quality and Applied Statistics
Rochester Institute of Technology
Rochester, New York

Marcel Dekker, Inc. New York and Basel
ASQC Quality Press Milwaukee

Library of Congress Cataloging-in-Publication Data

Barker, Thomas B.
Engineering quality by design: interpreting the Taguchi approach / Thomas B. Barker.
p. cm. -- (Statistics, textbooks and monographs; v. 113)
Includes bibliographical references and index.
ISBN 0-8247-8246-1 (alk. paper)
1. Quality control--Statistical methods. 2. Taguchi methods (Quality control). 3. Experimental design. I. Title. II. Series.
TS156.B3748 1990
658.5'62--dc20 90-3892
CIP

This book is printed on acid-free paper.

MARCEL DEKKER, INC.
270 Madison Avenue, New York, New York 10016

ASQC Quality Press
310 West Wisconsin Avenue, Milwaukee, Wisconsin 53203

Current printing (last digit):
10 9 8 7 6 5 4 3 2 1

PRINTED IN THE UNITED STATES OF AMERICA

Preface

The quality engineering by design methods that Dr. Genichi Taguchi introduced to the United States in 1980 have become one of the most talked about topics in the field of quality control and statistics. There are those who have utilized the techniques and made great quality and productivity gains but have not really understood what they have done. There are others who adamantly deny any benefit from the application of Taguchi's methods. Both blind application or outright repudiation are based on ignorance. It is easy to see why such ignorance abounds with this subject. The original work is found in two massive volumes only recently translated from the original Japanese. Some experts in the field of quality and especially experimental design have seen only bits and pieces taken out of context from Taguchi's philosophy and techniques and have misconstrued the intent of the application. Those who apply the methods do so based on short courses which concentrate only on Taguchi's methods and ignore the more general methods of quality and experimental design and analysis.

The purpose of this book is to shed some light on the lack of knowledge by "translating" the basic concepts of the quality engineering by design methodology into understandable words for the English-speaking world. It is also the intent of this book to fit Taguchi's methods in with the more general experimental design and analysis techniques and present an integrated view of a very powerful tool that builds quality into our products, processes, or services. For this reason the book has been written at a level that may be comprehended by working engineers (the audience targeted by Taguchi in the first place) and is also intended for quality professionals who wish to open their minds to a new application of experimental design.

Finally, I must thank the American Supplier Institute (ASI) for granting permission to reproduce and use the orthogonal arrays and linear graphs found throughout the book. The term Taguchi Methods™ is a trademark of ASI.

Thomas B. Barker

Contents

Engineering Quality by Design

1
The Concept of Quality

In the late 1970s and early 1980s a number of major changes began to take place in American industry. The automotive sector was in such a slump that thousands of workers lost their jobs in that industry. The ripple effect in other parts of the economy triggered a worldwide recession. One major automotive firm was rescued by a gigantic U. S. Government loan. It was in this same era that the Japanese had reached the point in their productivity and quality efforts that the fruits of their labors of over 25 years had begun to pay off. The Japanese learned that quality, productivity, and profits could peacefully coexist and actually reinforce each other. There have been television "white papers," magazine articles, and books suggesting and debating a connection between the above historical events. However, the fact remains that the balance of trade in the mid-1980s was the worst it has ever been in the history of the United States of America. The United States had become a debtor nation and a virtual colony of Japan.

In this same era, another change in the way major U. S. industries conducted business began to take place. This change was undoubtedly a reaction to slumping sales and profits. In some companies, it was probably a last ditch stand to guard against oblivion. It was also a change wrought with irony. As many companies began what was called *quality of work life* program (QWL), they simultaneously trimmed thousands of workers from their payrolls. The workers could not and did not believe in QWL. The employment reduction actions spoke louder than the QWL words. The word *quality* that had been popularly linked with a positive, good attribute became tainted. Management talked quality, but practiced just the opposite in the eyes of the workers.

Since the slogan QWL began with the word *quality,* quality professionals, who had been waiting in the wings for years, jumped on the new corporate bandwagons. The top quality consultants were welcomed with open arms by the CEOs. Top-down inspired quality improvement programs sprang up all over corporate America. Unfortunately, these programs were headed by managers who had never been associated with the methods of statistical quality control nor any other essential techniques that would be able to bring the levels of quality and productivity back into their companies. Most of these managers thought that quality was just another assignment, another program, and more a slogan than any real effort. Company-wide training programs were instituted. Some focused on the human aspects of quality. Others concentrated on the statistics—complete with all the supporting math. The result of this training brought wallpaper control charts, quality circle conferences, and slogans on slick posters. There was little improvement in quality and the balance of trade became worse. Quality awareness was rampant, but quality progress was slow to nonexistent.

The quality programs, the training courses, were only treating the symptoms. They were aimed at the last step of the process—when the product was manufactured. The research and development (R&D) groups were excluded from quality control (QC) activities. By ignoring the root causes of poor American quality in our efforts to improve quality, the task was set back to ground zero. Control charts were mere sugar pills in the medicine chest of quality cures. We needed a shot of penicillin to combat the bacteria that had grown into our system of developing and manufacturing products.

Enter Dr. Genichi Taguchi, a native of Japan—the very country that had helped precipitate some of the problems. He made his first visit to the United States in the summer of 1980, "to assist American industry improve the quality of their products. In a small way this was to repay the help the U.S. had given Japan after the war" (1). This visit was sponsored by a grant from the Aoyamagakuin University, and he made his base of operations the Quality Assurance Center of AT & T Bell Laboratories, Homdel, NJ. During this time, he visited, among others, Xerox and Lawrence Institute of Technology (LIT). He had a different message, and because of language barriers, he was almost ignored. According to Dr. John D. Hromi, then Chairman of Mechanical Engineering at LIT, "my students could not make head or tail out of what Taguchi was saying" (2). Back at Bell Labs, however, a young statistical engineer, Madhav Phadke, saw the potential in the philosophy and the methods and began to apply these methods with Dr. Taguchi's help (3). Two years later, both Ford Motor Company and Xerox recognized the potential of the Taguchi method and in 1983 began a vigorous educational program to promote the application among their suppliers and within their companies.

What was so appealing about the Taguchi method that attracted these top companies to make an investment in what has been called a radical approach to quality, experimental design, and engineering? (4). The answer was simple. Taguchi had presented a complete system of quality control that started with the product concept, extended through the design and engineering stages, and then into the manufacturing operation. No other quality advocate had presented such a comprehensive system—a system that has both the philosophy and tools to implement this philosophy. Taguchi's methods are not just references to standard, known statistical QC techniques like control charts or acceptance sampling plans. His engineering quality by design (EQD) techniques are an engineering-oriented, practical body of methods for making decisions about engineering design conditions. Unlike probability-based statistical methods, Taguchi's engineering-based concepts work in the deterministic environment of the engineer and the dollars and cents atmosphere of the business community. His philosophy begins and is founded on a new concept of quality. Instead of defining quality as a "good" attribute, he defines quality as:

QUALITY IS THE FINANCIAL LOSS TO SOCIETY AFTER THE ARTICLE IS SHIPPED (5).

In a sense, this definition depicts "unquality," since a loss to society is not a desirable characteristic.

The popular idea of quality is something beautiful and new and good. Look in your local telephone directory under *Q* for an example of such a concept. There are over 30 listings for *quality* in the Rochester, NY phone book including:

Quality Bakery; Quality Automotive; Quality Janitorial Supplies; Quality Beef; Quality Discount Office Furniture; Quality Forever, Inc.!!!

Since the popular concept of quality is conformance to a positive attribute, we will turn Taguchi's definition around while still retaining the basic concept. Quality in this light is:

QUALITY IS THE AVOIDANCE OF FINANCIAL LOSS TO SOCIETY AFTER THE ARTICLE IS SHIPPED.

In either definition, the important point lies in the fact that quality is related to a monetary loss, not to a gut feel or other emotional conditions.

To further emphasize this point, a quantitative relationship is used to link the language of things with the language of money. This is exactly what Dr. Joseph Juran admonishes the engineer to do. Juran says, "... become bilingual—speak the language of 'things' and the language of money." (6)

Taguchi's concept of quality and the loss function to describe quality does exactly what Juran prescribes! While it is possible to utilize other forms of a loss function, Taguchi has found the following quadratic form to be a practical, workable function.

$$L(x) = k(x - m)^2, \qquad (1\text{-}1)$$

where L is the loss in dollars (money), m is the point at which the characteristic (thing) should be set, x is where the characteristic actually is set, and k is a constant that depends on the magnitude of the characteristic (thing) and the monetary unit involved.

The basic quadratic loss function is used if no other function based on data is available. Market research operations would be responsible to obtain such data and to construct specific loss functions. This has been done in some cases in Japan. However, if a specific loss function based on market research is not available, the quadratic loss function will provide the necessary information to accomplish our goals of quality improvement. Such functions are common in the field of economics (7) and are not unique to the Taguchi philosophy. The appendix to this chapter shows how Taguchi devised his version of the quadratic loss function.

The important consequence of such a function is the fact that the farther the product's characteristic varies from the "proper" values, the greater the loss. Further, this loss is a continuous function and not a sudden step or an abrupt cliff from which we fall when there is a deviation from the ideal. This consequence of the continuous loss function illustrates the point that merely making a product within the specification limits does not necessarily mean that the product is of good quality, since good quality is now defined as keeping the product characteristic on target with low variation.

Automotive engineers have written specification limits on the size of doors and frames. The workers interpret these limits to mean that any door or frame manufactured within these limits is OK. From a statistical tolerancing point of view, the specification limits mean how far the tails of the distribution may extend given that the center of the distribution is centered on the midpoint of the specification. If a subpopulation of the doors are built "in spec" at the lower limit of the blueprint and a similar subpopulation of the frames are built at the upper limit of the print, we will have a mismatch. The doors will not fit into the frames properly.

That's exactly what happens on an American automobile assembly line. To make the door fit into the frame, the manufacturer hires strong craftsmen to bend and fit the doors to the frame (8). These door fitters account for a portion of the cost differential between Japanese built cars and those built in the United States.

In Japan, the parts of the car are built to nominal size rather than just within specification. The Japanese parts fit when they come together. There is no need for husky workers with rubber hammers to coax the doors to fit into the frames. The concept of building to nominal (or the midpoint of the specification) is not just another way of tightening the specification or putting an undue burden on the manufacturing organization; it is the correct, statistical way of understanding the meaning behind the toleranced specification.

It is important to look at the entire picture, not just one, isolated part. If we make doors and think that we have "zero defects" because we always build to specification, we are not considering the customer of these doors. The customer assembles the doors to the frames to make the product. We must strive for "zero defects" in the finished product. Our process and the control over this process will assure "zero defects." With an integrated

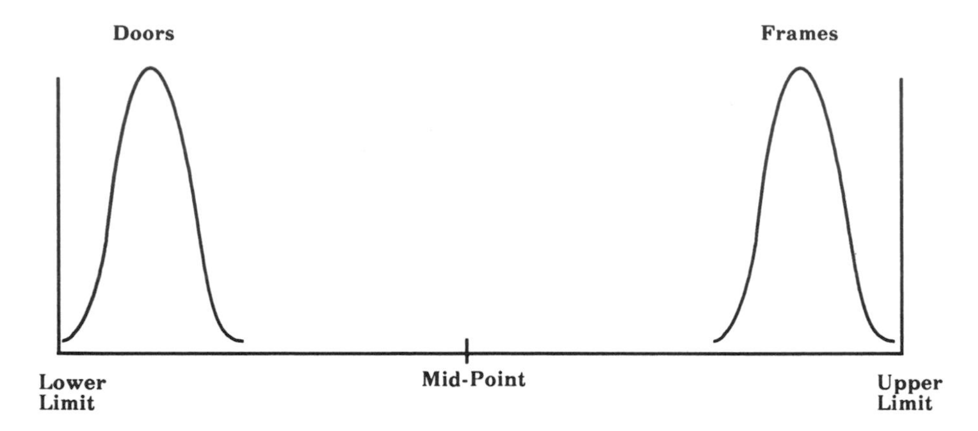

Figure 1-1.

view of the entire system, and with an understanding of the purpose of the specification, we *can* control. The loss function links this understanding to the hard facts of the monetary consequences of high variability and deviation from the targeted design point.

We will now look at some examples of the loss function taken from everyday life. By observing such losses, we can sharpen our awareness of this concept and begin to make use of the method more easily. (These examples are taken from the author's personal experience.)

POOR-FITTING SHOES

I have a difficult time buying shoes that fit properly. My size is 7D, which is not always stocked by the shoe store. Loss function metrics are best derived from market research. I have some limited data on my sensitivity to shoe size variation. I purchased a pair of size 7.5 dress shoes for $50 and found that even though the salesman assured me they would "break in" on my size 7 feet, they produced irritation and blisters on my heels. I ceased to wear them after only 2 days and consequently lost the value I had invested.

On the other end of the size spectrum, I bought a pair of size 6.5 white, summer shoes that were on final sale. They appeared to fit when I tried them on in the late winter (when my feet had experienced thermal contraction), but when I wore them (or tried to) in the heat of summer, the thermal expansion of my feet made an uncomfortable situation. These two sizes define the "drop dead" points for my shoe-fitting limits. Figure 1-2 illustrates this situation with the quadratic loss function.

With the three data points (including the comfortable size 7), it is possible to determine the constant (k) for the quadratic loss function and fill in the smooth curve, as shown in Figure 1-2. The constant (k) is determined as follows:

$$L(x) = k(x - m)^2,$$

$$k = \frac{L(x)}{(x - m)^2}.$$

$$k = \frac{\$50}{(6.5 - 7.0)^2},$$

$$k = \frac{\$50}{.25(\text{shoe size})^2},$$

$$k = \frac{\$200}{(\text{shoe size})^2}.$$

The magnitude of the constant (k) should not be an indication of the degree of the loss, since this constant is only a relative value for a particular situation. The size of the k depends on the monetary units used on the y axis and the measurement increment of the "thing" on the x axis. As the monetary unit gets larger, the k gets larger. As the measurement increment on the x axis gets smaller, the k becomes larger. We will now use the k value to produce the points in the smooth quadratic loss function over the range of shoe size, as shown in Figure 1-2.

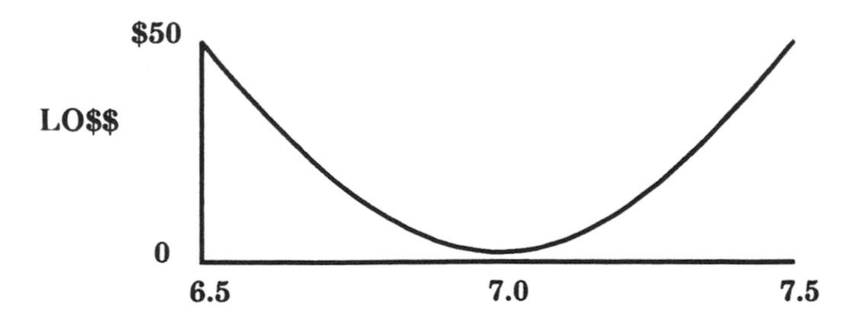

Figure 1-2.

Let's see what this loss function means in terms of the fit of the shoe. Imagine a shoe that is labeled *size 7,* but is in fact a 6⅞ (6.875) because the shoemaker skimps a bit on the leather. My feet will notice a certain tightness with such a shoe, and in the long run I will not wear this uncomfortable shoe as often as a perfectly fitting shoe. I will lose (according to the quadratic loss function) $3.12 of use due to the poor fit. I'm sure that you have poor-fitting clothes in your closet that follow the above concept.

There are, of course, other losses that could occur besides this fit problem. If the material in the shoe is not durable, there could be an early wear-out. We would construct another loss function for each quality characteristic. This is an important consideration in the total control of quality. Since we have a common (money) response, it is easier to make trade-offs between these characteristics based on the minimization of the total loss.

Returning to our shoe size example, we observe that as the size of the shoe deviates from the ideal size 7, the loss increases dramatically. The $3.12 loss (about the cost of lunch) at a deviation of one-eighth of a size unit is almost imperceptible, but if we deviate one-fourth or three-eighths from the ideal (7), the loss increases at a very fast rate due to the quadratic nature of the loss function.

In this homey example, I had to discard the shoes that were half a size too large or too small. My actual loss at either end of the spectrum was $50. This was a loss to me, personally, because I bought the wrong size shoes. I learned from this experience and now purchase only shoes that fit properly. I know my "drop dead" limits, but it cost me $100 to find them. Market research can be costly.

THE BRIEFCASE

This next example drawn from the author's experience shows the losses that poor quality can inflict upon the manufacturer. I received a new briefcase as a Christmas present in 1984. The case had a neat feature of a combination lock and a snap-shut latching mechanism. The latches were made of plastic and mated with the lightweight magnesium frame. When I received the case, the latches looked like the drawing in Figure 1-3a. After about a year of use, the right latch had worn down to the point that it would not hold the case closed and looked like the drawing in Figure 1-3b.

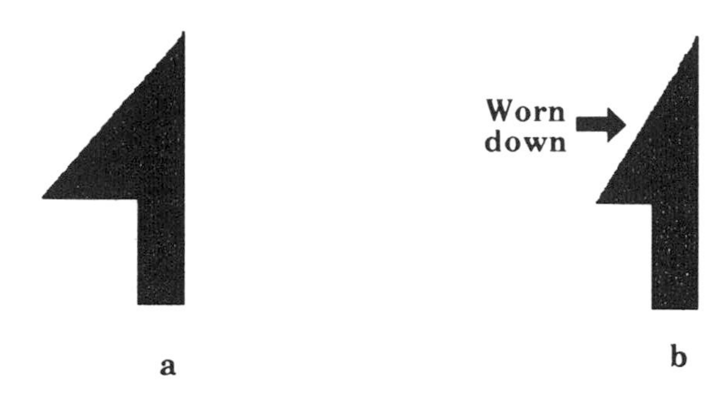

Figure 1-3.

I wrote to the manufacturer and included with my letter a close-up photograph of the worn latch and asked how to seek compensation under their 2-year warranty policy. After about a month without a reply, I called the company using my company's WATS line. I talked with a customer service representative, who gave me the names of two authorized service shops in my city. I chose the closest shop and took the case in for repair. It was a 10-mile drive each way and the repairman promised the job would be finished the following week. After another 20-mile round trip and a bill for $7.49, I had a functioning latch again.

A rivet had to be drilled out to replace the worn latch and there was a slight amount of damage to the interior of the case, since it had not been designed for serviceability. I asked the repairman what he had done with the worn-out part. He informed me that, like all the other ones he had replaced, he threw the broken one in the trash. I pressed him further on the number of cases he had repaired and, while not quoting an exact number, he gave me the impression that there was an epidemic of broken latches.

I sent the bill to the manufacturer as instructed and after about a month, I was compensated for the cost of the repair ($7.49). However, this bill was not the only cost. Let's add up the cost of the correspondence, telephone calls, travel, and all the other aspects of this warranty claim. Further, let's put the costs into two categories: my cost and the manufacturer's cost.

Item	My Overall Cost	Manufacturer's Cost
First letter	about $5.00	none
Reply	none	$20.00 (standard business letter cost)
Phone call	$3.00	none
40-mile trips	$8.00	none
Repair	reimbursed	$7.49
Totals	$16.00	$27.49

The original retail price of the case was only $69.95, so I doubt if there was any profit left after this repair. However, this is not the end of the story! About 3 months after the repair, I noticed the same pattern of wear taking place. Instead of taking the case back to the repair shop, I called the manufacturer and explained that this was the second time that the latch had worn down. The representative suggested that I send the case directly to the factory so they could fix it properly. I sent the case off to them on July 1, 1985 and did not see it again until October 16, 1985. In the months that I was without a case, there were so many calls, letters, and frustrations that I could have probably purchased another four cases! When I did get the case back, I observed that the factory had only done what the repair shop had done—a mere replacement of the latch. They had not sought out the cause of the problem. I carefully inspected the interface between the latch and the frame. I found that the frame opening into which the latch slid was very rough. This opening was acting like sandpaper and slowly abrading the plastic latch. I fixed the problem by removing the burrs from the frame opening and it has been operating properly ever since.

The above may seem to be a long story, but it illustrates a number of important quality-related lessons. The first is the quantification of losses. These losses are easily found by adding up the costs of doing business. Notice that both the manufacturer and the consumer accumulated losses, thus Taguchi's reference to "society" in his definition of loss. The second lesson is a common one. The worn part was replaced. We tend to simply replace parts rather than get to the cause of the problem. This particular part was wearing out prematurely, not because it was badly designed or of inferior materials. It wore out because another part was not finished properly. Instead of looking for the cause of the problem, the repairman as well as the factory repairperson simply replaced what was broken. There was a failure to look at the customer-vendor interface. The latch is the custom of the frame opening. To make a quality repair, we must find the source of the problem. We must ask *why* the latch failed.

The question, *why?*, is the most important question we can ask when a product, process, or service is not behaving in a quality fashion. I asked *why* the latch was wearing and found the source of the problem by observation (data). Often, though, the answer to the question *why* is more elusive than the simple rough edge of my case. Because of the complexity of many situations, we need a systematic method to ask *why*. This disciplined, systematic method that leads us to an understanding of our system is called *experimental design*. We shall see the methods of experimental design put to excellent use in the later chapters of this book.

The final lesson from this example concerns failure analysis. In the course of the many phone calls to the manufacturer, I managed to speak with the manager of quality assurance. He was not aware of the problem with the worn latches. It is no wonder he was unaware of the problem, since the repair stations were not returning the parts for analysis. It is not surprising that he did not understand why a foreign subsidiary of his company had redesigned the latch using metal rather than plastic.

Communication is a very important and critical aspect of the quality process. Every operating unit of a company from the cleaning staff to the chairman of the board must relay information into a central clearing house to ensure that quality problems are recognized and acted upon.

FEATURES AND QUALITY

While the popular concept of quality means a good attribute (and this is a difficult concept to modify), there is another concept of quality that has been ingrained into popular thinking by none other than Madison Avenue advertising campaigns. We have been lead to believe that the more features a product or service has attached to it, the higher the quality. If we inspect the definition of quality given to us by Taguchi, we see that there is no mention of anything beyond the statement of functionality. A feature is an extra attribute that embellishes the product beyond its basic function.

With a simple copying machine you must open the cover and place the original into position on the platen. A more feature-rich copier may include an automatic document handler (ADH) to feed the originals into position and then move them from the platen to allow the next original to be copied. This feature makes the job of copying a stack of documents easier, but it does not enhance the quality or functionality of the basic copier. If the ADH damages the originals or jams, the basic functionality is lost and quality disappears. Features usually add convenience to the operation of a basic device or enhance a service. However, they must not detract from the basic function of the product or service.

First-class seating on an airline does not hasten the trip or make travel safer, but it makes the trip more enjoyable. Cruise control for an automobile does not ensure the completion of the trip, but it makes driving less fatiguing. Features usually add cost to the item or service. A feature can be removed and the functionality will not be impaired. Features are necessary to gain a portion of the market and should not be eliminated from an option list, but they are not in the same category as quality until they fail to function properly.

When a feature ceases to operate it will impart a loss all by itself, but when it works it will not compensate for a loss due to the failure of a basic function of the system in which it is installed. Cruise control will not make the car go if the fuel pump has failed. We must consider only functionality in our measure of quality and use the loss function to quantify this functionality in monetary units.

LOSS FUNCTION APPLICATIONS FOR POPULATIONS

While the loss function is an excellent method of linking money and things, it is capable of going far beyond the mere description of loss via the graphics. We may use the loss function to actually compute the advantages of being on target with low variation for the distribution of a product characteristic.

In Figure 1-4, we see the relationship between output voltage and the gain of a power transistor in a regulated power supply circuit. This type of information, as well as the circuit diagram, is obtainable from transistor manuals published by the producers of electronic components.

If we have a specification of 115 V, it will be necessary to utilize a 20-gain transistor, which would cost 25 cents. The cost of such an electronic part depends upon the tolerance and the power-handling capability. The 25 cents quoted is for a tolerance of ±30%. While the word *tolerance* could have a variety of meanings to different individuals, we will interpret it to mean ± three standard deviations (SD) around the stated target value. Therefore, 1 SD for this transistor is 10%.

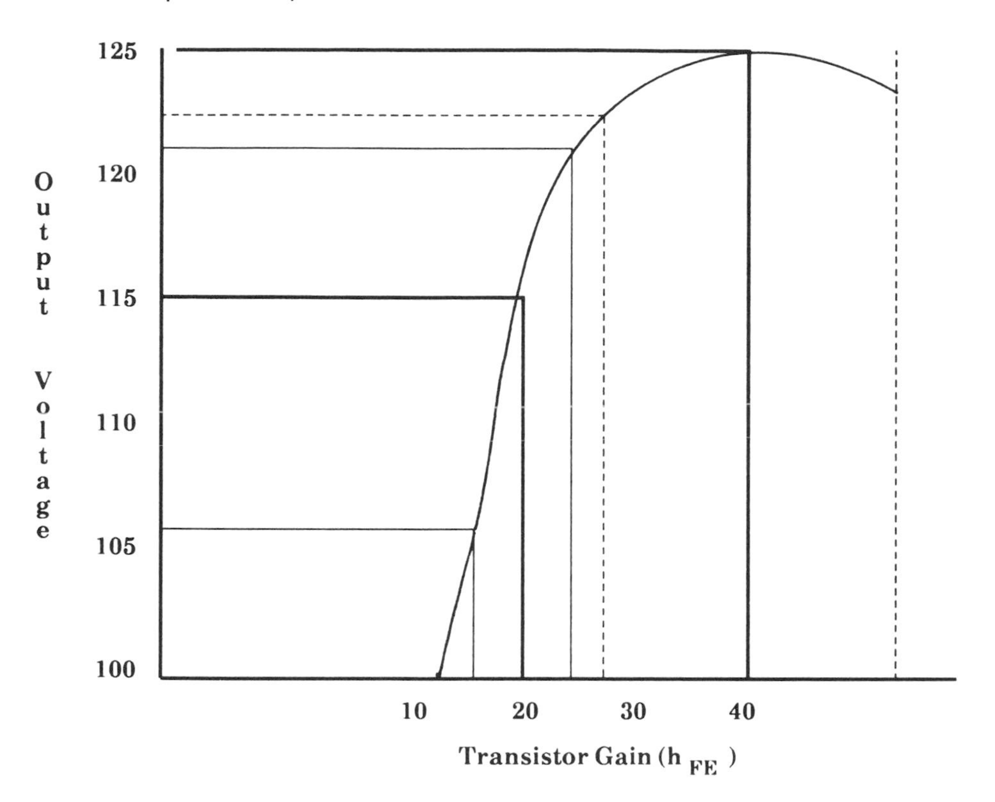

Figure 1-4.

When we superimpose the tolerance of the gain to the voltage vs. gain plot, we can see how the variation in the gain is transmitted to the variation in the voltage. Assuming a normal distribution of the gain, we will obtain a normal distribution of the voltage. While we are centered around the target of 115 V, it is possible to have a circuit with a voltage as low as 109 and as high as 121. We could have better control over the voltage by using the outlet plug in the wall! In such situations, it has been common practice in the United States to "throw money at the problem." A tighter tolerance transistor would be ordered in place of the 30% variety. While such a foolish idea gets the job done, it is very costly. Tightening the tolerance to produce a true regulated power supply could up-cost the design by a factor of four. A cost-effective, yet quality-conscious approach to this design problem would be to use that portion of the voltage vs. gain curve that is less steep. In this way, we do not transmit as much of the variation in the gain to the voltage variation. The portion of the curve where there is minimal transmission of variation occurs at the 40-h_{FE} point. Even with the low cost tolerance of ±30%, the variation in the voltage is only ±2 V when we work in the flat part of the voltage vs. gain relationship.

We will now show how the good engineering design utilized in the above example can lead to lower expected loss and, therefore, by Taguchi's definition of quality, a higher quality product. Figure 1-5 shows the loss function for deviations of the voltage from the target of 115.

We have superimposed the distribution functions for the two transistor choices on this loss function. Both will lead to high loss. The choice of the A transistor (20 h_{FE}) has a high loss because of its wide spread. The B transistor (40 h_{FE}) has a loss because it is off target. We must integrate the area of the loss function with the area of the distribution to calculate the expected loss. This may be done numerically, point by point, or by combining the distribution function with the loss function. It can be shown (see appendix) that this analytical combination will produce an expected

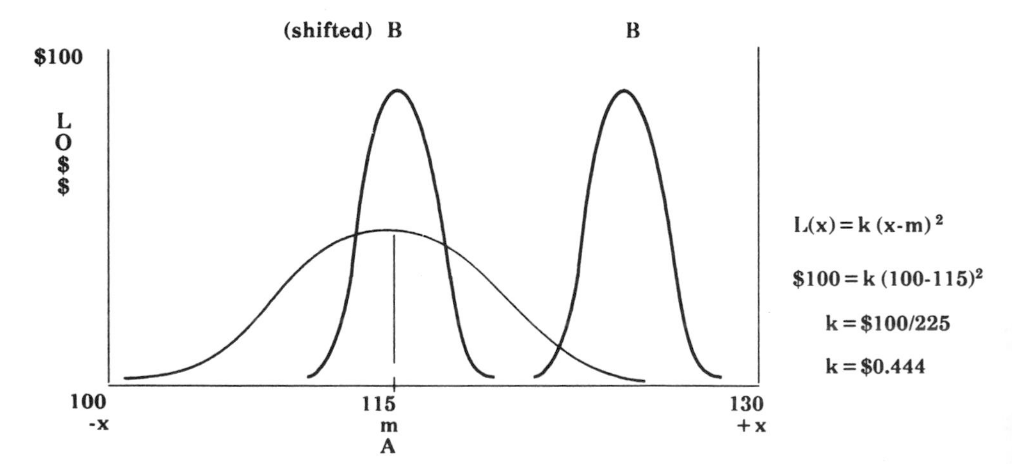

Figure 1-5.

loss (EL) that is related to "k," the standard deviation of the distribution (s) and the location of the average (AL) of the distribution with respect to the aim point (m).

$$EL = k[(AL - m)^2 + s^2]. \quad (1\text{-}2)$$

Therefore, for the two transistors, the losses are

For transistor A,

$$EL = .444[(115 - 115)^2 + 2^2].$$

$$EL = \$1.78.$$

For transistor B,

$$EL = .444[(124 - 115)^2 + .33^2].$$

$$EL = \$36.01.$$

The loss for transistor B is excessive, because its average level is 9 V off target. We may engineer around this problem by adjusting the value of another component in the circuit. The current limiting resistor recommended for the original circuit is 40 K ohms. However, we will modify this recommendation in our new design to 60 K ohms. By making this change, we are able to move the entire voltage vs. gain curve downward, as shown in Figure 1-6. This puts the B transistor's voltage on the target of 115 and the loss now becomes:

New loss for B,

$$EL = .444[(115 - 115)^2 + .33^2].$$

$$EL = \$0.048.$$

The above example illustrates the driving force behind Taguchi's quality engineering by design methodology. The expected loss drives the engineering design to minimize this loss. Loss is minimized when the product characteristic is on target with low variation. In the later chapters, we will see how it is possible to utilize this concept of minimization of loss by understanding the functional relationship between the factors we must control and the quality characteristic. We will gain this understanding by utilizing statistical experimental design.

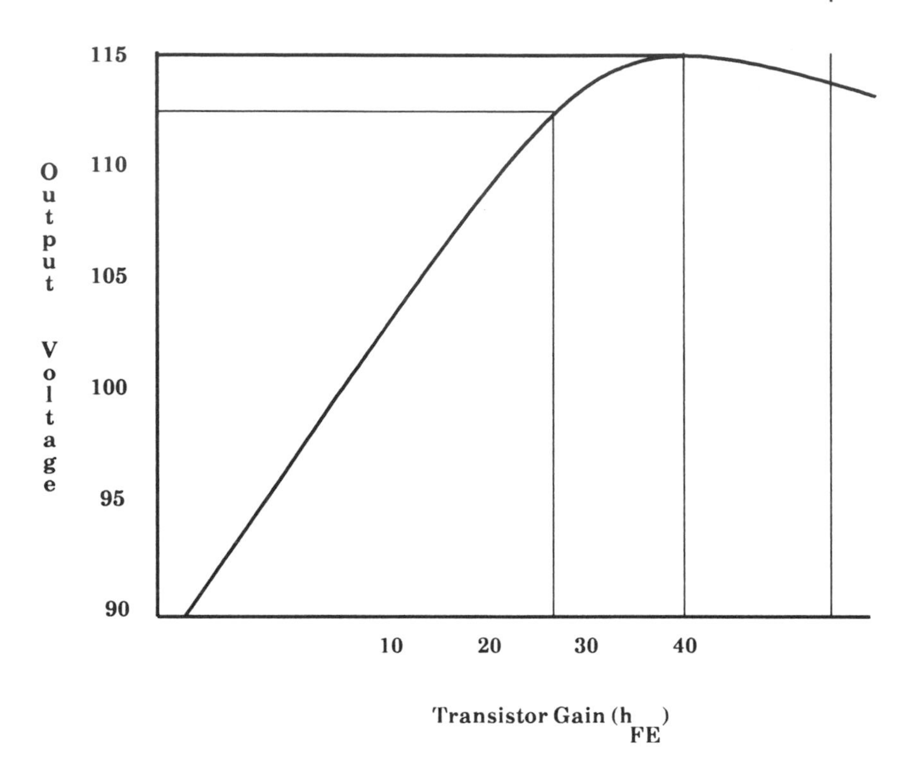

Figure 1-6.

REFERENCES

1. Dehnad, K., Ed. *Quality Control, Robust Design, and the Taguchi Method.* Wadsworth & Brooks/Cole, 1989.

2. Private communication, J. D. Hromi to T. B. Barker.

3. Phadke, M. S., Kackar, R. N., Speeney, D. V., and Grieco, M. J. Off-line Quality control for integrated circuit fabrication using experimental design. *Bell Syst. Tech. J.* 62, 1273- 1309, 1983.

4. How to make it right the first time. *Business Week,* June 8, 1987, pp. 142-143.

5. Wu. *Introduction Off-Line QC, Taguchi.* Central Japan Quality Control Association, 1979.

6. *Juran On Quality Improvement.* Juran Enterprises, New York, 1981.

7. Raiffa, H., Schalaifer, R. *Applied Statistical Decision Theory.* Harvard University Press, Cambridge, MA, 1961.

8. Sullivan, L. Reducing variability: A new approach to quality. *Qual. Prog.* July, 1984.

PROBLEMS FOR CHAPTER ONE

1. If the force exerted by a brush spring in an electric motor is too low (4 g/cm^2), armature contact will be lost and the motor will fail to operate. If the force is too high (16 g/cm^2), the brush will wear out prematurely and cause motor failure. The ideal force is 10 g/cm^2. The cost of a motor failure is the repair cost of $45.
   a) Compute the k value for this loss.
   b) Draw the quadratic loss function.
   c) Compute the expected loss for a production run of springs with an average force of 9.1 g/cm^2 and a SD of 1.58.
2. a) Identify three situations from your daily life activities that involved losses.
   b) Quantify the monetary losses associated with the above.
   c) Identify (or speculate) on the causes of these losses.
3. For one of the situations identified in question #2, construct a quadratic loss function (compute k and draw this function).
4. If the blackness (measures by ISO optical density) of the image from an electrostatic copy machine drops to 0.9, a service call is triggered at a cost of $70. The customer expects the density to be 1.5, which is the maximum attainable. Construct the loss function for this situation. (Hint: This is a single-sided loss function).
5. a) If the population of copier machines is able to hold an average density of 1.2 with a SD of 0.1, compute the expected loss using the loss function found in question #4.
   b) If the average density can be raised to 1.4 and the SD reduced to 0.033, what may the maximum cost of this "fix" or retrofit be if each copy machine experiences six service calls per year and the recapture period is 2 years?
6. The surface tension of an ink is critical to the proper operation of a printing press. The ideal surface tension is 20 dyn/cm^2. If the surface tension drops to 15 dyn/cm^2 the ink smears and there is lost production time (30 min at $50/hr) and lost materials (3.6 reams of paper at a cost of $6.23/ream). If the surface tension is too great (25 dyn/cm^2), the

press cloggs up and a 2-hr cleaning is required (at $50/hr).

a) Set up the two-sided, nonsymmetrical loss function.

b) What is the expected loss for an ink with an average of 20 dyn/cm^2 and a SD of 1.25 dyn/cm^2? (Hint: Compute the expected loss for one k value and take one-half of the result. Then compute the expected loss for the other k value and take one-half of this result. The final expected loss is the sum of these two results.)

7. (For advanced students)

The coefficient of friction of the brake device in a videotape recorder must be set at 0.6. If the friction deviates by more than 0.3 from this "ideal" setting in either direction, the recorder will malfunction. Too low a setting (0.3) will cause tape to spew into the recorder and cost $24 for a cleaning. If the friction is too high (0.9) the clutch is replaced at a cost of $72. Friction is related to the concentration of zinc sterate in the clutch material according to the following relationship:

$$\text{CoF} = e^{[-.18+(-.997\ \text{Ln}(x))]},$$

where CoF is the coefficient of friction, x is the percentage of zinc sterate, and Ln is the natural log.

a) Determine the two-sided, nonsymmetrical loss function for this situation.

b) Calculate the "ideal" zinc sterate concentration and determine the expected loss if the sterating process may be held to a SD of the following possibilities and costs:

SD zinc sterate	Control cost per clutch
.05	$.75
.10	$.45
.15	$.15

c) Devise a manufacturing/servicing strategy for this clutch.

APPENDIX 1-A

The Derivation of the Quadratic Loss Function (5).

Statement of identity:

$$L(x) = L(x). \qquad (1A\text{-}1)$$

Add m and subtract m on the right side of the above expression:

$$L(x) = L(m + x - m). \tag{1A-2}$$

A Taylor series expansion of 1A-2 produces:

$$L(x) = L(m) + \frac{L'(m)}{1!}(x - m) + \frac{L''(m)}{2!}(x - m)^2 + \cdots \tag{1A-3}$$

Since the loss is zero at m (our aim value), and the first derivative is also zero at m, we are left with the higher order terms. Because there is little added by terms greater than the quadratic (i.e., higher order terms are adding inconsequential harmonics) we are left with the following:

$$L(x) = \frac{L''(m)}{2!}(x - m)^2. \tag{1A-4}$$

The above (1A-4) expression may be simplified by making the derivative portion of the expression equal to a constant (k). This produces the familiar quadratic form of the loss function:

$$L(x) = k(x - m)^2. \tag{1A-5}$$

APPENDIX 1-B

Derivation of the Expected Loss (EL)

$$L(x) = k(x - m)^2. \tag{1B-1}$$

The expected loss (EL) over a distribution with mean μ is the average loss:

$$EL = k\frac{1}{n}\sum(x - m)^2. \tag{1B-2}$$

Add and subtract $\bar{x}$ on the right side of the expression:

$$EL = k\frac{1}{n}\sum[(x - \bar{x}) + (\bar{x} - m)]^2. \tag{1B-3}$$

Expand the square:

$$\sum(x - \bar{x})^2 + 2\sum(x - \bar{x})(\bar{x} - m) + \sum(\bar{x} - m)^2. \tag{1B-4}$$

The above expansion reduces to (since the $\sum(x - \bar{x}) = 0$):

$$EL = k\left[\frac{1}{n}\sum(x - \bar{x})^2 + \frac{1}{n}\sum(\bar{x} - m)^2\right]. \tag{1B-5}$$

However, $(1/n)\sum(x - \bar{x})^2$ is the variance, and since we are dealing with only one population with only one $\bar{x}$ for each expected loss calculation, we sum $(\bar{x} - m)$ only once (thus drop the summation sign and with $n = 1$, $1/n$ becomes merely 1), which results in

$$EL = k[(\bar{x} - m)^2 + s^2]. \tag{1B-6}$$

2
Concurrent Statistics

In the three stages devoted to the development of a product, process, or service, parameter design is the most important engineering activity. It is during this stage that we take advantage of the nature of the process to find the settings (or levels) of the factors (or as Taguchi called them, the *parameters*) that put our response at the correct value with low variation around this point.

Unfortunately, many processes do not exhibit low variation when they are put at the proper level of output, since many of our parameters influence both the level as well as the variation. These types of factors are called *control factors,* for they control the amount of variation in the product or process and may also have an influence on the level of output as well.

The transistor gain problem shown in Chapter 1 is an example of a control factor, since the voltage output level as well as the variation in voltage depend on the transistor gain set point. In that example, we had to find another factor to bring the voltage into the desired 115-V region after squeezing the variation out by taking advantage of the nonlinear nature of the voltage vs. gain relationship. The current limiting resistor was the component (or factor) that did the job of putting the voltage back to 115. Such a factor that has a major influence on the level, but very little if any influence on the variation, is called a *signal factor,* for it puts the signal (or level) at the proper position to satisfy the customer.

Many processes do not have the luxury of signal factors to recover from an off-target condition. In these situations, we must determine the trade-off between the location of the product characteristic and its variation. We need a summary statistic that incorporates both the average and the standard deviation in a single number. We have such a metric, which is called the *expected loss.* In the calculation of the expected loss there are essentially two components leading to the final loss. There is the loss due to the variation and the loss due to being off target.

Recall the formula for expected loss (EL):

$$EL = k\,[(\bar{x} - m)^2 + s^2]. \quad (2\text{-}1)$$

Table 2-1 shows how the summary statistic (EL) may be dissected into its component parts. Loss is a penalty for poor quality.

We showed with the transistor power supply example how this expected loss function is an effective figure of merit for making engineering design decisions. However, to use the expected loss function as a figure of merit, we must first obtain a loss function and a k value. In many circumstances, this is a very difficult task. A limited amount of market research is required to quantify the loss function starting points. Often this type of market research is very costly or simply impossible. Yet, we want to optimize the product characteristic for both the location value and the variation component. Taguchi recognized this dilemma early in the development of his methodologies and created a *transform for the function,* which he named the *signal to noise* (S/N). (It is interesting to note the origin of the name, *signal to noise.* Taguchi, a mechanical engineer, was working among a large number of electrical engineers at NTT. Not wishing to

Table 2-1.

EL	$\bar{x} - m$	s^2
SUMMARY PENALTY	PENALTY DUE TO BEING OFF TARGET	PENALTY DUE TO VARIATION

alienate them with new terminology, he picked a name for the figure of merit relating to optimal quality that his colleagues would accept. That name was signal to noise, a common figure of merit used in electrical engineering.)

The S/N is a concurrent statistic. A concurrent statistic is able to look at two characteristics of a distribution and roll these characteristics into a single number or figure of merit. A common concurrent statistic that is used in "classical" statistical analyses is the *coefficient of variation* (CV). The CV is simply the SD divided by the mean. It "normalizes" the degree of variation by the magnitude of the numbers being studied:

$$\text{coefficient of variation CV} = \frac{s}{\bar{X}}, \tag{2-2}$$

where s is the SD and $\bar{X}$ is the average.

As the variation around the average gets smaller, the CV decreases. Since smaller variation is desirable, this is the direction we would like to drive the CV. One embodiment of Taguchi's S/N figure of merit appears to be the inverse of the CV. While this form of the S/N is the least powerful of the variable data S/Ns, it is often utilized to introduce the concept of S/N as a figure of merit, because it appears in its formulation as signal (average) divided by noise (SD).

$$S/N = 20 \log_{10} \frac{\bar{X}}{s}, \tag{2-3}$$

where s is the SD and $\bar{X}$ is the average.

We shall briefly pursue the concept of S/N using the above expression, but then develop the true meaning of this quality metric with other more powerful and general transforms of the quadratic loss function.

Let us say we have the following set of values from an experiment that has been run under two conditions.

In Table 2-2 the mean values have remained the same from condition A to condition B, while the variation in condition B is much smaller. A smaller variation is a desirable characteristic and the resulting S/N ratio for condition B is larger. Since we want a large signal and a small noise, the S/N objective function is always optimized when it is made larger.

To obtain a feel for the relative difference between various S/Ns (which are measured in decibels [dB]), let us say that we have one process with a S/N of 23 and another process with a S/N 26. There is a 3-dB difference between these two processes. If we divide this 3-dB difference by

Table 2-2.

	CONDITION A	CONDITION B
	20	25
	25	35
	20	30
	40	28
	45	32
Average ($\overline{X}$)	30	30
Std. Dev. (s)	11.7	3.8
	S/N= 20 Log(30/11.7)	S/N= 20Log(30/3.8)
	S/N= 8.2 dB	S/N= 19.9 dB

10 (to return to bels), the difference is now 0.3, but 0.3 is a log (base 10) value. To put this into linear units (nonlog), we find the antilog of 0.3 by raising 10 (the base) to the 0.3 power. This produces the following:

$$10^{0.3} = 1.9952623.$$

When rounded, the above result becomes 2, indicating that a 3-dB change results in a doubling of the linear difference. Table 2-3 translates the decibel change into linear changes for selected gains.

From this table, we can see that it takes about a half a decibel gain to obtain about a 10% improvement in the response. Major changes (25–50%) begin to take place in the 1- to 2-dB region. It is important to understand the magnitude of the changes in a response variable, and Table 2-3 should help to accomplish this objective.

The S/N figure of merit that we have just defined is the simplest of a large number of such measures. All of these figures of merit are appropriately applied to help in the decision process that follows the design of experiments. The simple form of the S/N that we have just defined is an excellent introduction to the S/N concept, but, because it is very sensitive to changes in the variation as well as changes in the average level of our original response, it is not a very practical tool. Sometimes such sensitivities to both location and variation can mislead this form of the S/N.

Let us say that in a particular experiment we want to reduce the magnitude of a particular characteristic. If we use the S/N defined above

Table 2-3.

dB gain (change)	.1	.2	.3	.4	.5	1.0	2.0	3.0	6.0	10.0
Linear Ratio Change	1.023	1.047	1.072	1.096	1.122	1.259	1.585	2.0	4.0	10.0
Percent Change	2.3%	4.7%	7.2%	9.6%	12.2%	25.9%	58.5%	100%	400%	1000%

($\bar{X}$/s), we can see that the the S/N ratio will decrease as the average ($\bar{X}$) of the characteristic decreases! Since we always want the S/N to be large, the result for this situation goes in the "wrong" direction and leads to incorrect conclusions.

To combat incorrect conclusions, Dr. Taguchi has developed more than 70 different forms of the S/N ratio. The majority of these S/Ns are based on market research and are proprietary. However, we may utilize four specific, general forms for most continuous responses. These forms of the S/N will accommodate the following situations:

Table 2-4.

S/N "Type"	Desired Performance of the Original Response
S/N_S	REDUCE the value of the original response and at the same time reduce the variation. (Smaller original response is best)
S/N_B	INCREASE the value of the original response and at the same time reduce the variation. (Larger original response is best)
S/N_T	PUT response on a TARGET value and at the same time reduce the variation from this target. (Target original response is best)
S/N_N	KEEP response on a NOMINAL value while reducing the variation around this nominal response. (Nominal original response is best)

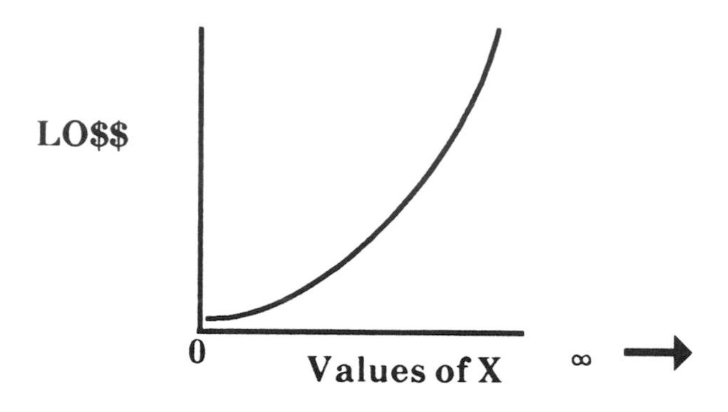

Figure 2-1.

We will now inspect each of the general forms of the S/N and observe its origin, its mathematical formula, and, most importantly, its relationship to the fundamental loss function. Figure 2-1 shows a single-sided, quadratic loss function with the minimum loss at zero.

As the values of X increase, the loss grows. Since we want low loss, we will target our X value to zero. The definition of loss (from Chapter 1) has the mathematical expression:

$$L(x) = k(x - m)^2. \tag{2-4}$$

Since the target (m) is zero, the above expression becomes

$$L(x) = k(x - 0)^2. \tag{2-5}$$

We may generalize the loss by using a constant (k) of 1 and also consider the expected value of the loss by summing all the losses for a population and dividing by the number of samples taken from this population. This produces the following expression:

$$EL = \left(\frac{\text{sum}(x^2)}{n}\right). \tag{2-6}$$

The above expression is a figure of demerit (loss). If we take the negative of this demerit expression, we have a positive quality function (utility). This is the thought process that goes into the creation of the S/N ratio from the basic quadratic loss function or, for that matter, any specific loss function that has a continuous relationship between loss and the original response variable. Taguchi adds the final touch to this transformed loss function by taking the log (base 10) of the negative expected loss and then he multiplies by 10 to put the metric into the "deci" "bel" terminology that was familiar to his electrical engineering colleagues at the NTT corporation. The final expression for the smaller is best S/N is

$$S/N_S = -10 \log_{10}\left(\frac{\text{sum}(x^2)}{n}\right). \qquad (2\text{-}7)$$

It is possible to follow the same thought pattern for the remaining S/N expressions (see exercises). The formulas are shown in Table 2-5.

There is a restriction on the values of the original responses that are used to generate the S/N_S. Since this figure of merit is based on a single-sided loss function, the values of the original responses may approach the value of zero but may not become less than zero. If the original response does take on negative values, we would lose the sense of the negative when the value is squared (i.e., $-2^2 = 4$; $2^2 = 4$). If we have an original response that attaches meaning to negative values, we will need to code the respon-

Table 2-5.

S/N "Type"	Loss Minimized When Original Response Is:	Mathematical Expression
S/N_S	REDUCED (Smaller is best)	$-10 \text{ Log } \frac{(\text{Sum } (x^2)}{n}$
S/N_B	INCREASED (Larger is best)	$-10 \text{ Log } \frac{(\text{Sum } (1/x^2)}{n}$
S/N_T	Put on TARGET (Target is best)	$10 \text{ Log } \frac{T^2}{s_T^2} \quad s_T^2 = \frac{\text{Sum}(x-T)^2}{n-1}$
S/N_N	Kept on NOMINAL (Nominal is best)	$10 \text{ Log } \frac{\bar{X}^2}{s^2} \quad s^2 = \frac{\text{Sum}(x-\bar{X})^2}{n-1}$

ses by adding a constant that will bring all of the numbers into the positive range.

TYPE B S/N

The type B S/N ratio is defined similarly to the type S S/N, but the x values are squared and then inverted. Again, we want to maximize the S/N_B, but in this situation we will do so by maximizing the original response. Figure 2-2 shows the single-sided quadratic loss function for a response that is bigger than the original response and for a better situation. The quadratic function described by this curve is $1/x^2$.

Figure 2-2 shows that as we reach toward infinity, the loss is minimized. Thus the implied target for the S/N_B is infinity (or a very large value of the response). Since we are finding the inverse of the x values in this calculation, the original responses may not become zero (the inverse of 0 is not defined mathematically). Also, negative values would lose their meaning, as we have already discussed with the type S S/N, and coding would be the correct procedure to remove negative numbers from the data set.

TYPE T S/N

When it is necessary to place our original response on a target value and losses will occur if the response deviates either lower than the target or higher than the target, we need to utilize the two-sided loss function and

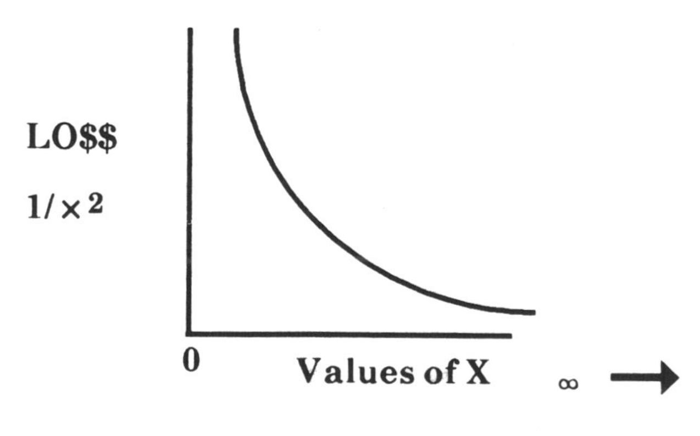

Figure 2-2.

the S/N transform, called the *target* S/N to noise. Again, if we go back to the quadratic loss function (expression 2-4), the value of m becomes T (for target) and the loss function may be written as follows:

$$L(x) = k(x - T)^2. \tag{2-8}$$

This loss function is the familiar double-sided form as shown in Figure 2-3. If the value of x deviates from the T value in either direction, we will have loss. Therefore, we wish to find the conditions that will put us on the target (T) and also keep the x values from wandering away from the target.

As the values of x deviate from the target, we experience greater and greater loss. This loss is quadratic by the definition of the loss function we are using. Since the S/N is a utility function (inverse of loss), we take the inverse of the loss as our figure of merit. However, this is not quite enough to ensure that we will seek and find the target value of our response. To ensure the fulfillment of this goal, multiply the inverse of the loss by the target value. This leads to the expression found in Table 2-5 for the type T S/N.

$$L(x) = k(x - m)^2. \tag{2-9}$$

We seek the target T to minimize the loss

$$L(x) = k(x - T)^2. \tag{2-10}$$

Using a generalized loss function with k=1 and also considering the expected (or average) loss over n items in the population, expression 2-10

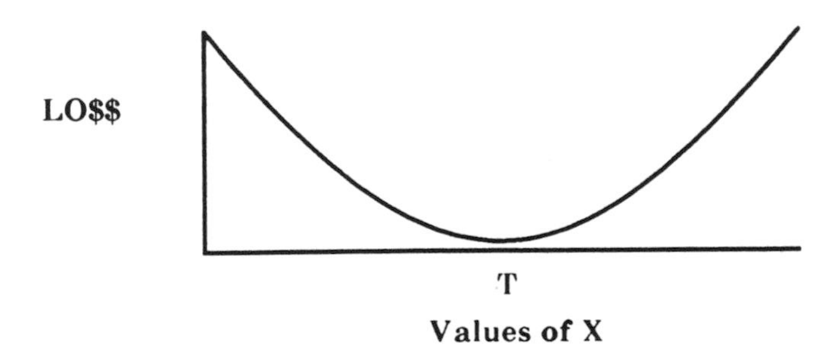

Figure 2-3.

becomes

$$EL = \frac{(x - T)^2}{n}. \tag{2-11}$$

We shall call this expected loss the S_T^2 (target variation). The inverse of the loss is the utility that, when multiplied by the target and then put into the decibel notation, becomes the S/N figure of merit:

$$S/N_T = 10 \log \frac{T^2}{S_T^2}. \tag{2-12}$$

The target type S/N is a very useful quality metric for finding the process conditions that will place our response on the target with small variations around this target. However, in the special case in which we have already found the target and it is our population average, this average has converged to the target and the S/N_T becomes the S/N_N that we originally used to introduce the concept in expression 2-3. Remember that we said that this form of the S/N is the weakest of the S/Ns. The reason for this weakness is the fact that we must have already found the process conditions that place our response on target. The usual recommendation with regard to "which S/N should I use" when it comes to a nominal or target situation is simply to use the type T (target) S/N, since it is a more general figure of merit and is not subject to confusion, as is the type N (nominal) S/N. Since the type N converges to the type T as the average converges to the target, you are always safe in using the type T.

EXAMPLES OF THE USE OF THE S/N

In rating the quality of photographic images, we may use a scale that ranges from 1 to 10, with 1 indicating a relatively poor image and 10 a relatively good image. The following data in Table 2-6 are taken from such an image-quality rating experiment.

We will take the eight observations from each of the eight experimental runs and compute the S/N (as well as the mean and the standard deviation). Since the judgment rule gives higher ratings to the better quality images, we will use the type B (bigger original response is best) S/N figure of merit. The summary statistics are in Table 2-7.

Notice how the S/N follows the mean value. In run #1, the lowest values are found and the S/N is also the lowest. Run #abc has the highest mean value and is also most consistent (has the lowest SD). High average and low standard deviation is exactly the combination of properties that

Table 2-6.

Experiment Identification	Observations 1	2	3	4	5	6	7	8
(1)	2	1	1	1	2	2	3	2
a	3	2	3	1	5	4	5	3
b	8	6	7	5	8	7	7	6
ab	8	5	9	7	9	8	10	8
c	3	2	2	2	4	3	3	1
ac	4	3	3	1	5	3	6	4
bc	7	5	6	5	7	5	5	5
abc	9	10	10	10	10	9	10	9

Table 2-7.

Experiment Identification	Observations 1	2	3	4	5	6	7	8	Average	Std. Dev.	S/N B
(1)	2	1	1	1	2	2	3	2	1.75	0.707	2.89
a	3	2	3	1	5	4	5	3	3.25	1.390	6.66
b	8	6	7	5	8	7	7	6	6.75	1.04	16.29
ab	8	5	9	7	9	8	10	8	8.00	1.51	17.51
c	3	2	2	2	4	3	3	1	2.50	0.93	5.71
ac	4	3	3	1	5	3	6	4	3.63	1.51	7.20
bc	7	5	6	5	7	5	5	5	5.63	0.92	14.74
abc	9	10	10	10	10	9	10	9	9.63	0.52	19.63

we are seeking for a low-loss, quality product. The S/N for abc is the highest (19.63) of the eight runs, confirming that this run is the best. There is a direct relationship between the avarage values and the type B S/N. As the averaged get larger, the S/N becomes larger. This is of course due to the fact that the greatest expected loss in a bigger-is-best situation comes from the deviation from the target and a lesser loss comes from the variation. As the values of the process come closer to the target, the variation component of the expected loss function becomes more of a driving force.

A WELDING PROCESS

In our next example of the use of the S/N, we will look at a welding process. In this process, we would like to avoid a depression or void at the weld interface. This void is measured in cubic millimeters. This is a clear case in which the smaller the characteristic is, the better off we will be, since our target value for the void volume is zero. The data from this experiment are found in Table 2-8.

Observe that the S/Ns have a negative sign. This may be thought of as a situation in which there is more noise than signal. This is exactly what is happening. The target value for this situation in which a smaller characteristic yields a better figure of merit is zero. Our data are all clustered over 100 units away from this zero point! Because of the negative sign, the S/Ns will become better when their absolute value becomes smaller. In Figure 2-4 observe this relationship on a number line.

The best S/N is found at run #4 with a value of −38.6. This run also has the smallest average void volume, but not the smallest standard deviation. The type S S/N, much like the type B S/N, is driven more by the average level than by the variation. However, as the original responses converge to the implied targets (zero for type S, infinity for type B), the impact of the level is diminished and the variation component takes over (see exercises).

The S/Ns that we just observed are based on single-sided quadratic loss functions. If we have a situation that produces loss because of

Table 2-8.

Run Identification	Observations 1	2	3	4	Average	Std.Dev.	S/N_S
1	100	95	125	85	101.3	17.0	-40.2
2	110	105	130	90	108.8	16.52	-40.8
3	125	115	139	99	119.5	16.84	-41.6
4	84	79	104	72	84.8	13.74	-38.6
5	92	85	110	80	91.8	13.12	-39.3
6	99	95	121	94	102.3	12.69	-40.2
7	99	96	120	87	100.5	13.96	-40.1
8	106	105	131	97	109.8	14.73	-40.9
9	118	117	140	105	120.0	14.58	-41.6

S/N:	-50	0	+50
	worse		better

Figure 2-4.

deviations in either direction from a midpoint or target value, we need to use the two-sided loss function. Table 2-9 shows data from an experiment taken from the automotive industry. In this situation the ideal bend angle is 65°. An experiment has been constructed to investigate variations on the bending process. Three methods were used and four bends were made from each method.

Besides calculating the familiar average ($\bar{X}$) and standard deviation (s) statistics, we have computed the type N S/N, the target deviation (s_T) and the type T S/N. Using the ordinary statistics ($\bar{X}$, s), we can see that there is little difference between the variation in these three methods, while method II comes closest to the specified (or target) value of 65°. Using our familiar statistics, it is very easy to pick the method that best suits our needs.

Table 2-9.

	Method I	Method II	Method III
	58.0	64.0	75.7
	57.0	65.5	76.5
	59.0	64.3	75.3
	59.0	63.0	74.1
$\bar{X}$:	58.3	64.2	75.4
s:	0.96	1.03	1.00
S/N_N:	35.7	35.9	37.5
s_T:	7.85	1.38	12.05
S/N_T:	18.4	33.5	14.6

Let us see if either of the S/Ns is able to duplicate this correct decision. If we look at the type N S/N, we see only a small change in the decibels over the three methods. This is partly due to the uniformity of the variation and partly due to the relative closeness of the average values. If we had to make a choice, we would pick method III with the S/N_N of 37.5, which is only 1.6 dB (about a 45% improvement) better than method II. This is not the correct choice. Statisticians would argue that the choice should be tested for *statistical significance,* which means that we would have to repeat the experiment a number of times to see how the S/Ns would vary. If the variation in the S/Ns exceeds the differences between the methods, then there is too much uncertainty to make a definitive decision and to pick one method over another. Sometimes such formal analyses (based on ANalysis of VAriance [ANOVA]) leads to an analysis paralysis and no decisions are made, ever! Taguchi no longer recommends the use of ANOVA and simply says "pick the highest value." This could lead to mistakes, *if* we use the wrong figure of merit.

This is exactly what has happened in this case. We tried to use the "nominal-is-best" S/N in a situation in which it was inappropriate to do so. In this situation, the average was changing. With a type N S/N, the average must be stable. We picked method III, because the average was largest in method III. The type N S/N is unfortunately driven by both the average level and the variation, and does not "know" what the aim level for our situation happens to be. It gets confused and we then will make mistakes in our decisions.

To guard against such mistakes in our decision process, the type T S/N was developed. This S/N is based on the two-sided, symmetrical quadratic loss function and uses the *target deviation* to drive the decision process. Note that the target deviations (s_T) in Table 2-9 show a wide range. This wide range is due to the fact that the target deviation is a function of the variation from the target and not a function of the variation from the average (variation from the average is the basis for the SD). We could look at the target variation as a figure of merit, but as it gets smaller it gets better. This is a demerit metric. So to keep consistency in our S/N concept (utility function), we take the inverse of the s_T and multiply this inverse by the target to anchor our S/N metric to the size of the response we are measuring.

The type T S/N easily selects method II by a spread of at least 15 dB (better than threefold improvement) over the next contender. This was an "easy one." The mean and standard deviation were able to perform the same feat. So why do we need a fancy S/N? We certainly do not want to overly complicate our decision process.

Let us look at another set of values that are less apt to be analyzed by the simple mean and standard deviation. Table 2-10 shows three more methods of making the angle. We come much closer to the correct target of 65° with these methods.

While method II is closest to the target, it has a large variation around this target. Method I is off target, but has half the variation of method II. Method III is off target by 2° and has a variation that is in between methods I and II. Without a loss function or type T S/N, this would be a difficult call to make. With these metrics, the job is easy. We find that the lowest loss is with method I, where we see a 1.2-dB gain over method II and a reduction of loss of 87 cents.

The S/N metrics are a set of concurrent statistics that are powerful tools in making quality-engineering decisions. We will begin to see how to use these tools in engineering designs by using statistical experimental design structures to efficiently determine the configurations of the methods we used above.

Table 2-10.

	Method I	Method II	Method III
	64.0	65.0	67.0
	63.5	64.0	66.0
	64.5	63.0	68.0
	63.0	66.0	66.0
	65.0	67.0	68.0
$\bar{X}$:	64.0	65.0	67.0
s:	0.79	1.58	1.00
S/N_T :	33.5	32.3	28.5
Expected Loss: (k= $1/degree)	$1.62	$2.49	$5.00

PROBLEMS FOR CHAPTER 2

1. In a similar fashion to that shown to relate the type S S/N to the single-sided quadratic loss function, show how the type B S/N is related to the single-sided, larger-is-better loss function.

2. In Table 2-8, the values of the void responses are measured in cubic millimeters. Convert the observations to cubic meters and recompute the type S S/N. Is there any difference in the relative ranking of the eight runs with this new metric? Why?
3. Compute the type S S/N for the following four sets of data. Compute the expected loss for each set given that k=1. Explain why the S/N does not change as much in the A – B comparison as it does in the C – D comparison.

A	B	C	D
90	95	.08	.09
90	95	.09	.095
100	100	.1	.1
110	105	.11	.105
110	105	.12	.11

4. Compute the type T S/N as well as the type N S/N for the following data. The target value for both sets is 900. Why is there a difference in the recommendations of these two S/Ns?

A	B
800	1050
850	1100
900	1200
950	1250
1000	1300

5. Another formulation used by Taguchi for the type N S/N when the standard deviation is equal to or greater than the average

$$S/N = 10 \log \left(\frac{\frac{1}{n}(S_m - s^2)}{s^2} \right).$$

$$S_m = \frac{(\text{sum } x)^2}{n},$$

where s^2 is the variance.
Show that when s^2 is smaller than the S_m, and therefore can be ignored in the numerator of the expression, this expression becomes expression 2-3 in the text.

3
Noise

While the signal-to-noise figure of merit provides a convenient tool for making the economic trade-offs between the variation and the position (location) of our fundamental quality metric, it is the recognition of the sources of variation (or *noise*) that is the key to successful quality engineering by design. Variation is the nemesis of quality, and Taguchi puts this nemesis into two categories. Inner noise comes from sources within our control and under our design authority. Outer noise is not under our control, cannot be delegated or legislated away, and usually raises its ugly head far after the final engineering design is frozen, or worse, when the product is in the hands of the customer.

Table 3-1 shows some examples of inner and outer noise and their consequences. Dimensional tolerances are inner noises and probably occupy more of the design engineer's effort and time than any other aspect of his or her discipline. Almost all engineering drawings have inscribed dimensions, and most of these dimensions have tolerances attached with them. As the first example of inner noise from Table 3-1 indicates, the fit of a piston within a cylinder of an internal combustion engine is critical. If the fit is too tight, friction will eat up part of the power stroke as well as the

Table 3-1.

INNER NOISE		OUTER NOISE	
Noise	Consequences	Noise	Consequences
• Dimensions (size) in an automobile engine piston and cylinder assembly	-too tight/loose a fit: Loss of horsepower -early wearout	• Atmospheric moisture	
		-Automobile	-low: poor economy -high:stalling
• Variation in transistor gain in a regulated power supply	-variation in voltage beyond customer requirements	-Copy machine	-low: low darkness -high:"dirty" copy
• Variation in charge voltage of an electrostatic copier's charge device	-Darkness variation -"Dirty" copies	• Public water supply's influence on fragrance of a shampoo	- Chemical reaction causes fragrance to become a stench

compression stroke and the design horsepower will fall short. If the fit is too loose, power will "blow by" the piston, also resulting in lower power. In either case there will be excessive wear. The tight fit rubs too much and the loose fit rattles or bounces on the cylinder walls. Besides poor short-time performance characteristics, the quality in the long run (reliability) has much to be desired. A great deal of effort is exerted to avoid mismatching pistons and cylinders. Countermeasures such as selective assembly (where computer-controlled matching of dimensions of pistons and cylinders is a part of the daily routine of engine assembly) overcomes such dimensional noise.

Figure 3-1 shows the relationship between the output voltage of a regulated power supply as a function of transistor gain (h_{FE}). The variation (noise) in the transistor ($\pm 30\%$) transmits variation to the voltage. There is a direct, almost amplified transmission of the transistor's variation to the voltage, which makes the "regulated" power supply resemble a variable voltage power supply if we compare one assembly to the next. Customers would certainly not want to do business with a vendor that was foisting such devices on them.

If the above power supply was being used in an electrostatic copier, the variation in output voltage would cause a wide variety of density

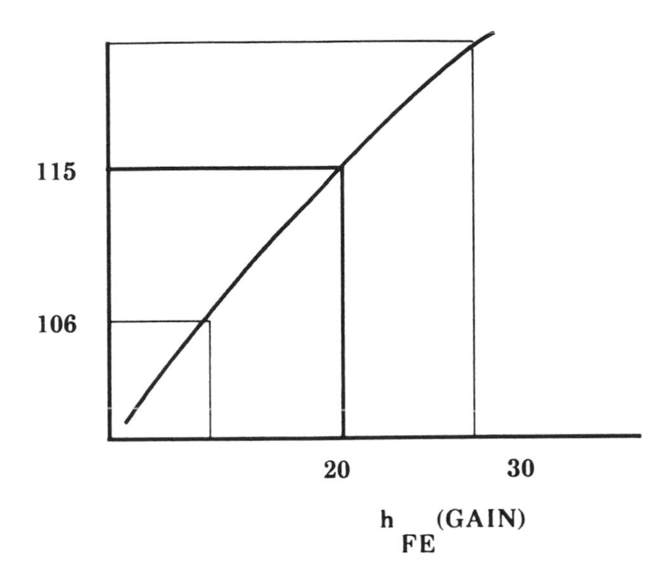

Figure 3-1.

changes and "dirty" copy (called background) conditions to exhibit themselves as defects. These defects would cause a loss to the copier manufacturer in the form of increased service calls and warranty costs, as well as a tarnished reputation (or *loss of face* in Japanese terminology).

Now, the above noises are examples that are within the control of the manufacturing process. The manufacturing engineer in the engine assembly plant can devise methods to measure, sort, and match pistons with cylinders. The electrical engineer can specify a higher grade (tighter tolerance) transistor for the power supply. The copier manufacturer can sort power supplies to find the right ones for the application. All of these "quality control" measures are costly. Such measures give quality the bad image of "expensive." We shall see in later chapters that quality does not have to be costly if the knowledge of the process is obtained through proper experimental methods.

While we are able to take control of the inner (or manufacturing noise) sources of variation, the noise outside of our control (outer noise) will occur as it pleases without warning or mercy. A sudden torrential rainstorm will strand dozens of cars on a saturated highway. The cause is ignition-wire shorting. A dry, cold winter will bring on reduced fuel economy in our automobiles. We have witnessed both of these manifestations of the uncontrollable outer noise of relative humidity (RH).

Electrostatic copy machines, because of the nature of electrostatics, experience copy quality swings caused by changes in RH. During the dry winters, the toner material that produces the image becomes more highly charged and refuses to transfer to the latent image. This causes a low darkness image, as shown in Figure 3-2a. At the other end of the moisture spectrum, the hot, damp summer months give rise to low electrostatic charge and the toner falls into non-image areas, giving rise to what is called *background,* as illustrated in Figure 3-2b.

A countermeasure to the background problem used in the early copier machines was a heater to dry the toner material while it was idle over-

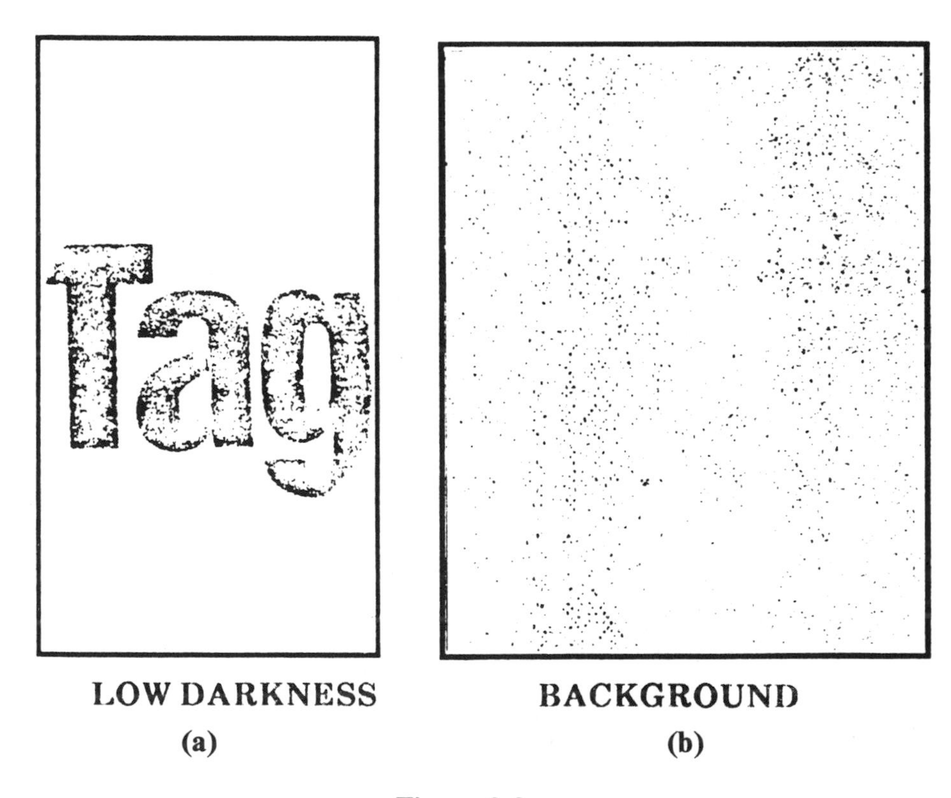

Figure 3-2.

night. This costly "fix" is about as good a remedy as using a pair of rubber gloves to fix a leaky fountain pen! The heater is an extra part prone to failure, and the real, inherent problem of moisture sensitivity has not been solved.

We will explore one last example of outer noise. Most noise-induced quality failures will lead to a loss function. In this instance, the mineral content of the water supply reacted unfavorably with the fragrance and other ingredients of the shampoo to produce a smell like burning rubber. The countermeasure was an investigation by the research and development personnel to find a formulation of the shampoo that would not react with the water unfavorably. This example shows how an outer noise factor caught the manufacturer after the product was on the shelves in the stores. Such design and product testing mistakes are very costly, both in terms of "face" and in terms of the cost of the investigation and subsequent corrective measures.

Now that we are able to distinguish and categorize noise into controllable (inner noise) and uncontrollable (outer noise), we are ready to embark on the journey that will take us into experimental design methods that will allow us to take control of both of these noises to produce products that are very close to being variation free.

PROBLEMS FOR CHAPTER 3

1. List at least three sources of outer noise that you have encountered in your ordinary, daily life. Indicate the consequences of these noises.
2. Select three sources of outer noise from your work environment. What countermeasures were used to overcome these noises?
3. List three sources of inner noise from your work experience and the consequences of these noises. How did your company overcome these noises?

4

Taking Aim at Noise— An Arsenal of Experimental Tools

Up to this point we have merely stated the case for improved quality and have shown ways (loss function, S/N) to measure that improvement. However, comparative measurements of product characteristics cannot be made unless we are able to change the characteristic. While the normal process of producing a product will cause the characteristic to vary, it is important for quality engineers to have the ability to trace the sources of variation back to their root causes.

Such systematic methods of being able to assign cause have been available since the mid-1930s and are known as the methods of *experimental design.* These methods had their roots in the field of agriculture and were developed in England by Sir Ronald Fisher at the Rothampstead Agricultural station outside of London. Before such methods were available, trial and error was the main method of gaining information about a new process or product. Even today, over a half century since their invention, experimental design methods are not used as widely as current quality problems might dictate, since very few engineers have learned about them. This lack of knowledge may be attributed to the apparent complexity of these methods, along with their statistical analysis. It has only been in the past few years, with the advent of simple-to-use computer programs, that experimental design has begun to emerge as a powerful tool. We may also thank Dr. Taguchi for enhancing this tool via his simplification and cataloging methods. It is no longer necessary to hold a degree in statistics to utilize the power of experimental design. An engineer with a minimal amount of instruction may immediately begin to apply experimental design to real problems in need of the understanding that will lead to a solution and the high-quality product dictated by the market.

Design engineers and process engineers have been testing for eons. Even in the ancient era of the craftsman whose only specification was "make more of the same," a test of the final assembled product was an important aspect of the product delivery process. Of course, if the product did not behave as the craftsman thought it should, he or she would rework it until it functioned in the manner promised to the customer. In a way, product engineering has not progressed much beyond this simple testing. When a new design is put into a prototype configuration, or a new process is fired up for the first time in a process capability study, the old tried-and-true testing method is used. A test of this type is of course necessary to confirm our expectations of the product or process (just like it was for the craftsman), but our processes are so complex that often the test fails to confirm our expectations. When this happens, we often make a minor change in the conditions, based on our limited experience, and try again. This trial-and-error method often involves only one factor being changed at a time, since changing any more at once would confuse the issue. We probably learned this method of changing one factor at a time early in our careers or even in college physics or chemistry labs. Changing one factor at a time is a dangerous habit that grows on us, since it has the psychological effect of immediate feedback. Unfortunately, the feedback is often negative.

Engineers are often astonished that it is possible to change more than one factor at a time and to not get confused. It is the methodology of *statistical experimental design* (SED), which allows multiple factor experimentation. SED, coupled with statistical analysis, is able to "go one further" and separate the information from the variation that often clouds our decisions. Let's look at a rather simple experimental design and observe how a one factor at a time (1-Faat) would handle the problem and then see how SED approaches it.

Our example will involve three factors. A factor is usually a general design condition that we are investigating for a proper setting or level. There are often many factors in a particular engineering design. The levels of the design factor are often purposely changed in an experiment so as to find the best conditions that will lead to a product that meets the customer's requirements. Table 4-1 lists the factors and the levels for this example.

We will measure the image quality of a printed object that has light and dark density, ragged and smooth edges, and narrow and wide lines. Image quality may be measured by asking the customer his or her opinion of the image. This opinion may be obtained by techniques called *psychometric methods.* It is not important to know the details of these methods to observe and understand the differences between the experimental techniques. Let us assume that we are able to obtain a quantitative, precise, and meaningful response for the problem.

Our goal is to find the conditions that will maximize the image quality (IQ). We think that IQ is influenced by the three factors listed in Table 4-1, so we will test this belief out and at the same time find the best conditions that maximize IQ. In the 1-Faat "experiment," we simply change each factor from a base case (usually the low level for all factors) to another level or condition. Table 4-2 shows the 1-Faat's four runs and the responses to these conditions.

The analysis for this experiment is quite easy. We simply find the difference between the first run and each of the others and assign this effect to the factor being studied. Table 4-3 shows this calculation.

From the above differences, we can see that only the change in sharpness and the change in line width had an effect on the IQ. The best conditions would involve a sharp, wide image of either light or dark density.

While the above 1-Faat seems to produce the correct combinations that lead to a proper set of design conditions for the IQ problem, there is another method, called a *two-level factorial,* that we should investigate. Now, the word *factorial* does not have the meaning associated with the mathematical notation "!" (e.g., 5! means $5 \cdot 4 \cdot 3 \cdot 2 \cdot 1 = 120$), but merely

Table 4-1.

Factor	level(-)	level(+)
Density	light	dark
Raggedness	ragged	sharp
Width	narrow	wide

Table 4-2.

Run	Density	Raggedness	Width	IQ
1	light	ragged	narrow	.77
2	dark	ragged	narrow	.77
3	light	sharp	narrow	1.43
4	light	ragged	wide	1.17

Table 4-3.

Runs being compared	difference observed	conclusion
2(dark) .77 1(BASE) .77	0	Dark images have no effect
3(sharp)1.43 1(BASE) .77	.66	A sharp image is better
4(wide) 1.17 1(BASE) .77	.40	A wide image is better

means that we will be looking at *factors.* Table 4-4 shows the experimental layout for our IQ problem. Notice that more than one factor changes at a time in many of the runs.

The analysis of this design is not very difficult. To illustrate the ease of drawing conclusions, we will take each factor, starting with the width, and find the contrast between the two conditions or levels. We start with the width, since it is so easy to see the four conditions grouped together. It is also easy to see that in each group of four values for the width, the same changes are taking place with the other two factors. This observation we

Table 4-4.

Density	Raggedness	Width	IQ
light	rough	narrow	.48
dark	rough	narrow	.79
light	sharp	narrow	.62
dark	sharp	narrow	1.58
light	rough	wide	.92
dark	rough	wide	.99
light	sharp	wide	1.32
dark	sharp	wide	2.24

have just made is called *balance* and is one of the reasons that statistical experimental design methods are able to find the information we need with a minimal expenditure of resources. Table 4-5 shows the method of contrasting the narrow lines with the wide lines.

The difference between these two averages is 0.5. We can find similar differences between the averages of the light and dark densities and the rough and smooth images. Table 4-6 shows a summary of these differences.

Table 4-5.

	narrow	wide	
	.48	.92	
	.79	.99	
	.62	1.32	
	1.58	2.24	
Total:	3.47	5.47	
Average:	.87	1.37	Difference= 0.5

Table 4-6.

Factor	Average Difference Between levels
Density	.57
Raggedness	.65
Width	.50

Each of the differences in Table 4-6 is based on averages of four observations, rather than differences between single observations as in the 1-Faat method. It is possible to average over other factors, since we have the balance in the statistical design; that is, whatever is happening due to all the other factors at one level of the factor being studied is also happening in the same way due to all the other factors at the other level. Because of this balance, we are able to piggyback analyses without confusing one effect with another. This absence of confusion (or *confounding,* which is the proper statistical word) between factors leads to another property of the statistical experimental design, called *orthogonality.* Orthogonality means that we may study each factor independently. This does not mean that the factors are acting independently in a physical sense, but that the mathematics of the analysis is unconfounded. If there is a physical dependence between the factors and how they influence the response, then we have the second reason for the factorial experimental design. To see how this works, let's plot the data for each factor, see Figure 4-1.

To get an even greater insight into the message these data are trying to deliver, we will plot a family of lines showing the change in the IQ as a function of density, but over distinct levels of raggedness. This is a differential plot and indicates if there is a different behavior for the density effect based on its physical dependence on the raggedness factor. If we see a differential in the slopes of the lines, we shall declare that these two factors *interact* and we know that they are not independent in a physical sense. Figure 4-2 shows this interaction.

From this interaction plot we can see that there is indeed a difference in the change in IQ due to density, depending on the level of the raggedness. With rough (unsharp) images, the change in quality from low density to high density is relatively minor compared with the IQ change due to

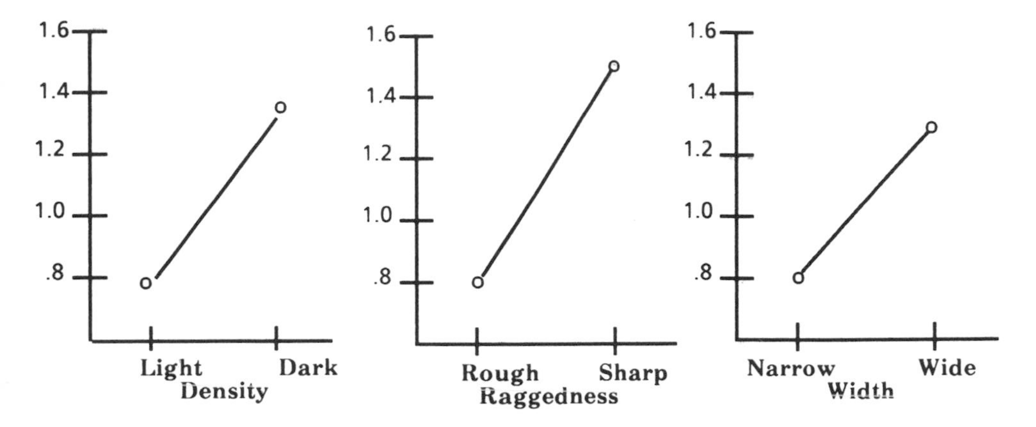

Figure 4-1.

density when the image is sharp. In fact, when we combine the preferred high density with the preferred sharp image, we obtain a level of IQ that is higher than we would have expected by just looking at each factor singly, as we did in the 1-Faat testing method. What we have just observed is an interaction between density and raggedness, which is superadditive. Such superadditive interactions are quite important and form the basis of many physical and chemical reactions. Another such interaction occurs between two chemical agents used to develop the film you drop off at a neighborhood 1-hour photo shop. Alone, these developing agents react slowly, but in combination they superadd to allow the speedy delivery of our pictures, to which we have become accustomed.

The ability of the factorial experimental design to detect interactions is the most important reason for its utilization. Add this to the ability to find average effects without extra runs, and we may begin to understand why this type of experimental method forms the basis for efficient experimentation.

The factorial type of experiment is the cornerstone of the Taguchi technique of infusing quality into a basic product design or process. However, if we were to use just the simple method of testing all possible combinations of all factors at all levels to solve our problems, we would soon find that, with even a reasonable number of factors, there are just too many combinations! This is where the methods of experimental design become just a bit more complicated and where we truly utilize the sam-

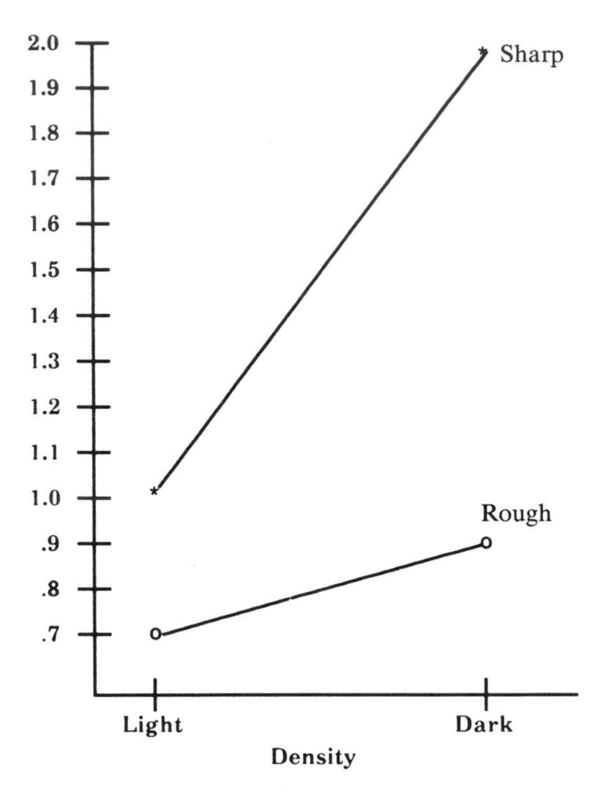

Figure 4-2.

pling power of statistical experimental design. Instead of looking at all the possible combinations of factors and levels, we take a sample from the population. Unlike samples taken at random in the usual statistical methods, the samples taken by statistical experimental design are specific and lead to the correct conclusions via trend analysis, much like the trends we observed with the IQ experiment. Let's look at the IQ experiment once again to determine its structure. We had three factors: density, raggedness, and line width. Each factor had two levels or conditions associated with it. The number of runs (or treatment combinations) came to eight. We may generalize the notation for a two-level experiment as follows:

$$k = \text{number of factors.}$$

$$2^k = \text{the number of runs.} \tag{4.1}$$

Therefore, if we have three factors (as we did) then there will be

$$2^3 = 8 \text{ runs.}$$

We may compute the number of runs for two-level designs with various numbers of factors as follows:

Table 4-7 shows that as the number of factors grows, the number of runs increases very rapidly. While we may be able to live with this condition for simple experiments that do not require a long amount of time to complete, or are sparce in factors (six or fewer), the type of problems we often encounter are complex and have many factors. Therefore, the simple 2^k factorial does not meet our requirements for practicality. Rather than using the 2^k method, we will use an extension of this method, called the 2^{k-p} fractional factorial experiment.

While the initial impetus for the fractional factorial design is based on reducing the amount of work, there is still another more compelling reason to use it. In experiments with many factors, there are many possibilities of interactions taking place. We have seen that an interaction is an important piece of information. However, while interactions are important, the occurrence of interactions, especially interactions between more than two factors, is not prevalent. This is an observation that has been made by many experimenters and follows the Pareto principle, in which a vital few items cause the chief effects and there are many trivial effects left over. An even more important observation concerning the types of effects taking place in experiments concerns the complexity of the in-

Table 4-7.

# factors (k)	# runs 2^k
2	4
3	8
4	16
5	32
6	64
7	128
8	256
9	512
10	1024

teractions. While two-factor interactions are common, three-factor interactions and higher order interactions, for all practical purposes, are extinct. When we run all of the possible combinations in the full factorial experiments, we make a provision in the design to be able to detect all of these interactions. Every time we make a provision to detect a piece of information, we must do work and expend resources. Expending resources, when it leads to "empty" information, is not efficient.

We have spoken about information in a rather informal manner up to now. We must have a way of counting the information in our experiments and a convenient counter is called *degrees of freedom* (df). Very simply stated, df represent the independent information obtainable from experimentation. An example of df would involve weighing five objects and finding the average.

Object #	Weight
1	120
2	140
3	110
4	160
5	70
Total	600

Average = 600/5 = 120

Given that we have the average of 120 and we were to freely choose 4 of the 5 original objects, it would be impossible to select a different object as the fifth object while still maintaining the average of 120. The name *degrees of freedom* comes from the number of free choices we have. Before we computed the average, we had five free choices of objects, but after the average was computed, we were only able to choose four objects from the group freely, the last object depended upon the average we had computed. The five original pieces of information were summarized in the average value, which in itself is a piece of information. Information cannot be created. Information comes from work. The work that produces information comes from our experiments. if we do five pieces of work (the weighting of the objects is work), then we may never have more than five pieces of information. If we compute an average, we have converted one of the original pieces of information into this average, which is dependent upon the values of the five objects.

The (total) number of degrees of freedom (df) available is equal to the number of observations (n), minus one.

$$df = n - 1. \tag{4-2}$$

Degrees of freedom relate to our experiments as follows. Since we are looking for average effects, then the degrees of freedom for each factor shall be one less than the number of levels for that factor. For a two-level design, there will be 1 df for each factor. If we have a three-level design, there will be 2 df for each factor, etc.

Interactions have a special rule based on the same information theory. The df for interactions is computed by multiplying the df of the interacting factors together. If we have 2 df for factor A and 2 df for factor B, then the df of the A · B interactions will be 4. In two - level designs, since each factor has 1 df, all interactions have 1 df. Having only 1 df is a very important consideration in this type of experimental design. We will see this importance when we investigate three-level designs.

Having this formal method of counting information, let's look at the information content of some typical two-level experiments in Table 4-8.

As we can see, as the number of factors increases, the degrees of freedom (which cost us resources) increases very quickly. To remain efficient, we will take some of the "empty" information made up of the higher order interactions and purposely confuse (or confound in statistical language) it with information that has meaning. To illustrate this point, let's redo the IQ experiment with a fourth factor, the amount of light available for viewing the images. In Taguchi's terminology this would be a *noise* factor.

Table 4-8.

number of factors k	number of single effects	number of 2-factor interactions	number of 3-factor interactions	number of 4-factor interactions
5	5	10	10	5
6	6	15	20	15
7	7	21	35	35
8	8	28	56	70

With four factors there should be 16 runs. Table 4-9 has only eight runs. How did we do it? Let's look at the generic, bare-bones structure of the original (first) three-factor IQ experiment. We will use a minus sign to indicate the first level and a plus sign to indicate the second level. We will also use letters of the alphabet to indicate the factors. The design then becomes:

It is possible to find the following interactions from this experiment:

$$A \cdot B$$
$$A \cdot C$$
$$B \cdot C$$
$$A \cdot B \cdot C$$

Table 4-9.

Density	Raggedness	Width	Light
lo	rough	narrow	dark
hi	rough	narrow	bright
lo	smooth	narrow	bright
hi	smooth	narrow	dark
lo	rough	wide	bright
hi	rough	wide	dark
lo	smooth	wide	dark
hi	smooth	wide	bright

Table 4-10.

A	B	C
−	−	−
+	−	−
−	+	−
+	+	−
−	−	+
+	−	+
−	+	+
+	+	+

We might expect an A · B interaction, as well as the A · C and B · C interactions, but the A · B · C interaction (a three-factor interaction) is usually not expected. Therefore it is empty in information content and we may apply it for something more useful, such as another factor, e.g., the amount of light available to view the images. The A · B · C interaction has a set of *levels* designated by the minus and plus signs. We obtain these signs by doing exactly what the expression for the interaction tells us to do. In each row we multiply the sign of each factor times the sign of the other factors in the interaction expression. (Note the " · " means to multiply). In the first row, we multiply the "−" times another "−" times another "−" to produce a "−". We continue to do this same operation for the remaining seven rows. This would lead to a new column in our design for A · B · C, and the matrix looks like this:

Table 4-11.

A	B	C	A·B·C
−	−	−	−
+	−	−	+
−	+	−	+
+	+	−	−
−	−	+	+
+	−	+	−
−	+	+	−
+	+	+	+

The A · B · C interaction is usually not expected information, so we will utilize it as another factor, D. Now our design matrix becomes:

Table 4-12.

A	B	C	D
−	−	−	−
+	−	−	+
−	+	−	+
+	+	−	−
−	−	+	+
+	−	+	−
−	+	+	−
+	+	+	+

Thus, to add another factor to a design without expanding the size of the experiment, we take a less useful piece of information and confound it with a more useful piece of information. This is the basic principle of the fractional factorial design and applies to designs with more than merely two levels. Let's look at the full factorial (Table 4-13) that provided the base for the sample found in Table 4.12. Those runs that are in the smaller design (Table 4-12) are marked with "[ ]".

The systematic sample from the full factorial design, based on a selection process tied to the "empty" interactions, is the method of generating fractional factorial design arrays or *orthogonal arrays,* as Taguchi calls them. However, as the number of levels expands beyond two, the complexity of the process is greatly increased. This is where Dr. Taguchi has made some nice simplifications. We will now take the next step and look at the general consequences of using fractional factorial designs from the traditional, statistical viewpoint and then show how Taguchi's methods simplify these traditional approaches.

When we built the fractional factorial design, we purposely made factor D take upon itself the conditions of the ABC interaction. In doing so, we lost the ability to study the ABC interaction, but gained the ability to study another factor (D). But what about the remaining two-factor interactions that we might want to study? Let's look at the design again, but this time put in the other interaction columns. We calculate these interaction columns in the same way that we calculate the ABC interaction column, by multiplying the signs in each row by the factors in the interaction. Thus, the AB interaction is the result of the multiplication of the sign in the A column times the sign in the B column. In Table 4-14, the first row, this is a "−" times a "−", which produces a "+." For the next row, we have a "+" in the A column and a "−" in the B column, so we obtain a "−" in the A · B interaction column. The procedure is repeated for the remaining rows of this AB interaction and for the remaining interaction columns.

We have computed the interaction columns, but have not put them in alphabetical order. Rather, the order used shows the similarity of the signs in the columns in three distinct groupings. The AB and CD interactions show the same sign pattern. The AC and BD, as well as the AD and BC, show the same patterns in two more groups. The analysis of the interaction effects is done in the same way as we do the analysis of the single effects; that is, we find the average of the responses at the "+" condition and the average of the responses at the "−" condition and contrast these differences to see if anything is taking place due to this effect. Now the contrast for AB and CD will be exactly the same, because the sign pattern is the same for these two effects. This is a physical picture of confounding and is the price we pay for cutting the size of the experiment down to a

Table 4-13.

A	B	C	D
[−	−	−	−]
+	−	−	−
−	+	−	−
[+	+	−	−]
−	−	+	−
[+	−	+	−]
[−	+	+	−]
+	+	+	−
−	−	−	+
[+	−	−	+]
[−	+	−	+]
+	+	−	+
[−	−	+	+]
+	−	+	+
−	+	+	+
[+	+	+	+]

Table 4-14.

A	B	C	D=ABC	AB	CD	AC	BD	AD	BC
−	−	−	−	+	+	+	+	+	+
+	−	−	+	−	−	−	−	+	+
−	+	−	+	−	−	+	+	−	−
+	+	−	−	+	+	−	−	−	−
−	−	+	+	+	+	−	−	−	−
+	−	+	−	−	−	+	+	−	−
−	+	+	−	−	−	−	−	+	+
+	+	+	+	+	+	+	+	+	+

more manageable set of runs. We can see a similar problem for the AC-BD interaction and the AD-BC interaction. We are able to measure only the combined effect of these interactions with our fractional factorial design.

We could work out this pattern of confounding each time we design an experiment, but since the pattern will remain the same, we may work it out once and then use a general structure to build our particular experiment. Table 4-15 has the same structure as Table 4-14, but we have added column numbers to identify the seven pieces of information available from the eight runs. Notice that this is consistent with the information counting rules for df; that is, we did eight pieces of work (eight runs) and therefore the total df is equal to eight minus one, the seven columns that give seven independent pieces of information.

The statistical approach to defining this general pattern of confounding requires the use of a branch of mathematics called *group theory.* However, we may take the results of this theory and build a simplified nomographic system for the information needed to construct our fractional factorials. Dr. Taguchi has developed this system (and was presented with a Demming Award for it), which he calls *linear graphs.*

To use linear graphs, we simply take the column numbers in the design, as shown in Table 4-15, with the linear graph shown below in Figure 4-3. The columns numbers shown at the ends of the lines represent the single factors, and the interaction column numbers are along the line.

Therefore, the interaction information for the factors in columns 1 and 2 will appear in column 5. The interaction information for the factors in columns 2 and 3 will show up in column 7. The interaction information between the factors in columns 1 and 3 will appear in column 6. It is not

Table 4-15.

column:	1	2	3	4	5		6		7	
	A	B	C	D=ABC	AB	CD	AC	BD	AD	BC
	−	−	−	−	+	+	+	+	+	+
	+	−	−	+	−	−	−	−	+	+
	−	+	−	+	−	−	+	+	−	−
	+	+	−	−	+	+	−	−	−	−
	−	−	+	+	+	+	−	−	−	−
	+	−	+	−	−	−	+	+	−	−
	−	+	+	−	−	−	−	−	+	+
	+	+	+	+	+	+	+	+	+	+

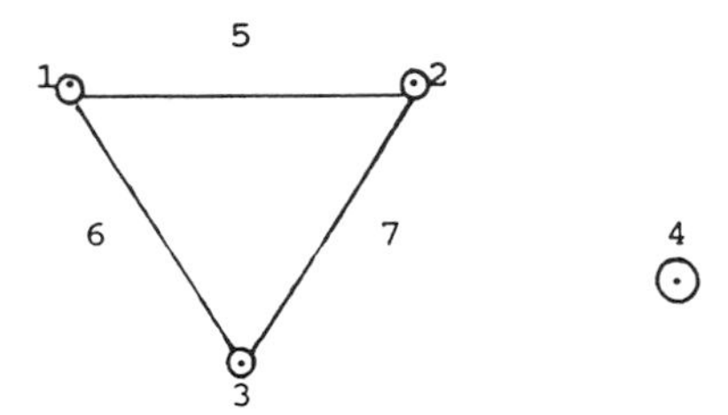

Figure 4-3.

possible to display the interaction between all three factors or any interaction beyond a two-factor interaction with a linear graph, so the fourth column is placed off to the side and can be used as another factor in the experiment, as we already demonstrated.

So what does the linear graph tell us? Simply stated, it is an early warning device that helps us to prevent likely single effects from becoming confounded with likely two-factor interactions. For instance, if we expected the AB interaction, we would avoid placing another single factor in the fifth column of this design where the AB interaction resides. However, if we would not expect an AB interaction, then we could use the fifth column to investigate another factor. Such a design would look like this.

While the linear graph method is a quick, easy way to avoid the hazards of confounding likely effects with each other, it does not give the entire pattern of confounding in the experimental structure. We have

Table 4-16.

column:	1	2	3	4	5
factor:	A	B	C	D=ABC	E=AB
	−	−	−	−	+
	+	−	−	+	−
	−	+	−	+	−
	+	+	−	−	+
	−	−	+	+	+
	+	−	+	−	−
	−	+	+	−	−
	+	+	+	+	+

already seen that the AB interaction is confounded with the CD interaction, so in the above design, factor E would also be confounded with CD. There are many other confounding patterns that are only demonstrated by the defining relationships derived from the group theory starting point, as mentioned previously. To fully understand the nature of the confounding in fractional factorials, you must utilize the defining relationships, which are covered in most books on experimental design. For the application of the *Taguchi techniques,* we will be able to utilize the linear graphs as the necessary guides to minimizing the confounding between factors and interactions.

THREE-LEVEL DESIGNS

If we expect a curved relationship between the factors under study and the response, we will need at least three levels. The same basic principles of fractional factorial design structure that we learned for two-level designs apply to three-level experiments. The idea is to take an empty piece of information (such as a three-factor interaction or above) and combine (or confound) it with another factor that we need to study. The only problem with experiments with more than two levels is that the information requirement (as measured by df) grows very quickly. As the information requirement grows, the amount of work grows rapidly. In a two-level design, each factor has but 1 df and all the interactions have only 1 df. However in a three-level design, each factor has 2 df and the two-factor interactions have 4 df (2 df · 2 df)! The statistical basis for the construction of three-level designs is very complex and is poorly treated in most experimental design textbooks. Again, we will utilize Dr. Taguchi's linear graphs to ensure that there is as little confounding in our designs as possible, without having to learn another level of complex theory.

The design in Table 4-17 is the smallest of the three-level configurations. It can basically investigate two factors if there are interactions expected but could, in the absence of any interactions, look at up to four factors. We can, of course, obtain information on the linear effects as well as the curved relationships in the form of quadratics, and this is the main reason for the use of three-level designs.

The linear graph shows that if a factor is placed in the first column and another factor is placed in the second column, then the interaction information (remember we need 4 df) between these factors will fall into the third and fourth columns. the 4 df of the interaction has been split into two separate, 2 df columns to allow two more, three-level factors to be placed

Table 4-17.

Column:	1	2	3	4
	A	B	AB	AB^2
	1	1	1	1
	2	1	2	2
	3	1	3	3
	1	2	2	3
	2	2	3	1
	3	2	1	2
	1	3	3	2
	2	3	1	3
	3	3	2	1

Linear Graph:

1.________.2
3,4

Note: The levels are designated by 1,2,3 rather than "-" and "+" for this 3-level configuration.

into the experiment, if there is no expected interaction between any of the factors. Since it is likely that interactions will exist, it is dangerous to use this design for more than two factors.

Now that we have seen how the fractional factorial designs are constructed and how they can be configured to investigate the factors that influence quality, we will put them to work in the next chapters. The tables and strategies found in Chapter 10 show the useful two- and three-level designs and include comments on the operational aspects of these design structures. We will show these designs in action in the parameter and tolerance designs and thereby obtain a practical working ability with these statistical sampling techniques that produce the information we need, while expending the least resources.

REFERENCE

Barker, T. B. *Quality by Experimental Design.* Marcel Dekker, New York, 1985.

PROBLEMS FOR CHAPTER 4

1. A chemical engineer has modified a process to increase the throughput. The major quality characteristic has unfortunately suffered by this change. The process team has identified the following factors, levels, and interactions. There is only 1 month (20 working days) available to conduct an experiment. If it takes 1 day to run each test in such an experiment, set up the appropriate orthogonal array to solve this problem.

Factor	Levels	
Temperature	100°C	200°C
Pressure	24 psi	36 psi
Time	5 min	15 min
Concentration	.5%	.75%
Batch size	1 l	10 l
Raw material	Lot #23	Lot #37

The possible interactions are: pressure-temperature, concentration-temperature concentration-time, concentration-raw material, and batch size-time.

2. An automotive engineering team is given the task of increasing the fleet average miles per gallon by improving the efficiency of a V-8 engine. The following factors have been identified during a brainstorming session as possible contributors to changing the number of miles per gallon of fuel:

Factors	Working range
Engine displacement	4–6 l
Piston stroke	Long stroke–short stroke
Block type	Iron–aluminum
Compression ratio	8.5–10.5
Valve timing	35–40° BTDC
Fuel flow	1–15 lbs/hr
Spark advance	10–20° BTDC
Fuel octane	87–93
Type of spark plug	L type–Circle type
Speed	20–55 mph
Atmospheric moisture (relative humidity)	20–90% RH

Propose an experimental design that will elevate the most influential factors from this list for inclusion in a second-phase experiment. Justify your recommendations.

3. The experiment you proposed in problem #2 was run and the following factors were identified and selected for the second phase. Propose a three-level experiment for this investigation. The first three factors are more likely to interact with each other than the last four factors.

Influential factors	Working range
Compression ratio	8.5–10.5
Valve timing	35–40° BTDC
Fuel flow	1–15 lbs/hr
Spark advance	10–20° BTDC
Fuel octane	87–93
Speed	20–55 mph
Atmospheric moisture (relative humidity)	20–90% RH

4. A photographic engineer has been told to increase the perceived image quality and to do so at the least cost. He determines that making a formulation change to the developed solution would be the quickest and least costly approach to this requirement. His team has selected the following ingredients and process characteristics for the experiment:

Ingredient	Working range
Metol	4–8 g/l
Hydroquinone	8–16 g/l
Alkali	20–40 g/l
Sulfite	20–40 g/l
KBr	1–3 g/l

Process parameter	Working range
Temperature	70–80° F
Speed of travel	5–8 in/sec

The team expects that there will be some curvilinear relationships between the IQ rating and the factors. They know that there is an interaction between the metol and the hydroquinone. Propose and lay out a possible experimental design configuration for this situation.

APPENDIX
CHAPTER 4
ORGANIZING FOR EXPERIMENTATION

While the experimental strategies we have developed in this chapter will lead us to better engineering designs, these structures by themselves will fail to deliver optimal results without good engineering input. Experimental design is a joint venture between the engineering team's prior knowledge and the focusing of this knowledge via experimental design configurations.

It is essential that the prior engineering knowledge is well thought out and organized before an experimental structure is constructed. The following is an approach to organization for experimentation that has produced quite successful results in the past.

First, Assemble the Team

Successful experimentation relies on a team effort. The mix of team members should include the following.

1. *The experts*—That group of engineers that has guided the development of the product, process, or service through its infancy and to the point that the engineering quality by design approach has the potential of optimizing their efforts.
2. *Operators or technicians*—Those responsible for hands-on-contact with the product design, process development, or service. These people will also be responsible for conducting the experiment.
3. *Semiexperts*—Knowledgeable engineers who are not working on this particular development project, but know enough to ask embarrassing questions. Also known as *peripheral experts.*
4. *The customer*—The group or person who will directly receive the output of the experimental results.

It is important that the above critical mass be assembled to brainstorm the problem. If the experts are the only ones allowed to work on the problem, they will probably have the answer all worked out in advance and not see a need to experiment. By having the semiexperts ask questions, the experts' blinders are stripped away so that they have a wider vision of the scope of the problem, and the first obvious answer is no longer the only possible solution.

The operators and technicians are intimate with the process. They may even know more than the experts! Respect and utilize their inputs. Also, since these people will be running the actual experimental tests, be sure that they know that some of the runs will produce poor responses. Often, without the type of involvement we are suggesting, the technicians will fix the experiment up to make sure that the results come out "right"! Remember, a good experiment does not always make good products, but it will make good information. Another aspect of employee involvement regarding the technical force is to present the results of the experiment to this group first. Don't make the first presentation to management! There are two good reasons for this strategy. First, you involve the technical staff. After all it is their experiment and they have a right to know the results and the ramifications of these results. Second, and probably more importantly, if you have made any blunders in reporting the information from the experiment, the technical staff will be the first to help you get it right. Better them than your boss!

The final part of the team that you assemble for the brainstorming session is the customer. Again, involvement is a prime reason to invite the customer who comes with the vision of helping to build an experiment that will impact his or her own activities. Give the customer ownership of the experiment and you will not have to sell the customer on the results of the experiment. The other reason for customer involvement is the fact that the customer is best suited to identify the response variables you will measure as a result of the experimental runs. Who knows best what is important to measure? Of course it is the customer, who will probably be measuring every item you deliver in either a formal or informal manner. You should be measuring your experimental results using the same criteria used by the customer and only the customer is able to tell you those criteria.

Involvement and information are the keys to assembling the brainstorming team. If there are others in your organization who would fall into these categories, make them a part of the team.

To what size should an experimental team grow? If there is one representative from each group, then the smallest team would have four members and could work. A dozen participants is closer to the ideal and is a very good size. As the team grows, however, the brainstorming session loses some of its spontaneity and is harder for the leader to handle. Teams as large as 30 participants have functioned but are the exception, just as are the small four-member teams.

Scheduling the Organizational Meetings

There are at least two meetings required to organize experimentation. The first meeting is a brainstorming session and the second meeting is a follow-up to rationalize the many thoughts that emerge from brainstorming. Scheduling these meetings is dictated by the 5-day work week and work customs. Mondays are considered poor thinking days and Fridays are often taken as holidays. There should be no more than 2 days between the brainstorming meeting and the follow-up meeting. This leaves us with Tuesday for brainstorming and Thursday for the follow-up. Do not try to do the job of brainstorming and follow-up in 1 day. It does not work! The brainstorming session should be scheduled first thing in the morning and only 1 hr is required for this exercise. The follow-up requires half a day. The following agendas are offered as guides to organizing these meetings.

BRAINSTORMING MEETING

Tuesday, 8:30 AM to 9:30 AM

Agenda

Introduction and statement of problem	10 min
Brainstorming	20 min
Factor type identification	30 min

FOLLOW-UP TO BRAINSTORMING MEETING

Thursday, 8:00 AM to 12 Noon

Agenda

Review of goals	10 min
Review of brainstorming ideas —any additions —purge the ridiculous ideas —find redundancies and consolidate	50 min
Elevate the critical factors for consideration in the experiment	60 min
Select the factors for the experiment	30 min
Identify levels for the factors	30 min
Determine possible interactions between the factors in the experiment	45 min
Review action items and assign the timetable for the statistical activities	15 min

Conducting the Brainstorming Session

Having assembled the critical mix of participants, we will now look at an actual brainstorming meeting. The leader of the session should have the following equipment:

1. Flip chart
2. Black, green, and red heavy marking pens
3. A roll of masking tape (to tape filled pages on the wall)

The brainstorming session is run much like any typical brainstorming activity. We will not cover the basics of brainstorming in-depth here. This is a skill that should have been acquired in other quality training programs. However, remember to strive for quantity; let any idea no matter how "dumb" emerge; write the ideas down in large print and display previous ideas by taping each filled flip-chart page on the wall of the room to help catalyze more ideas.

There are some particular enhancements to the brainstorming process that we will include in its application to experimental design. It is important to be able to identify three types of factors during brainstorming. Most participants will recognize the need to identify the factors that influence the process and are under the control of the experimenters. These are known as *controlling factors* and will probably make up the majority of the list. However, we must also identify the response variables that change as a function of the controlling factors. There will probably be a prime response variable that has precipitated the need for the experiment in the first place. There will also be other response variables that must be identified, because as we make an improvement to the prime response, we may cause another response to degrade. We will need to balance the experiment so that we do not fix one problem and create another. The third type of factor is the outside noise that we spoke of in Chapter 3. These factors must be considered since we will want to design our product, process, or service to be immune to variation in such outside noises. To confirm this immunity, we must place such outside noise factors in the experiment and we must identify these factors for such an inclusion. To summarize, the brainstorming must identify three types of factors:

1. Controlling factors
2. Response variables
3. Outside noise factors

During the brainstorming process, the leader should encourage the participants to cover all of the above. After the actual brainstorming is finished (only about 20 min!) the remainder of the meeting is devoted to sorting the types of factors identified. A green marker may be used to identify the response variables. (Place a large *R* next to each response). A red marker may be used to identify the noise factors. (Place a large *N* next to each noise factor.) The remaining factors are by default the controlling factors. If there is any time left, the team may begin to help the leader sort the factors into categories.

For example, such categories in a mechanical process may include dimensional and frictional characteristics. In a chemical process there may be formulation and process categories. In an electrical process there may be current, resistance, and voltage categories. Sorting into categories helps find the redundancies by placing similar factor names together. Since during a brainstorming session more than 100 ideas may emerge, it is possible that similar ideas will come up more than once. Categorizing helps narrow the list.

Between the Tuesday brainstorming meeting and the Thursday rationalization session, the leader will compile the organized ideas from the flip-chart pages and get the information to each of the participants by no later than 2:00 P.M. on the Tuesday of the meeting. This action allows the participants to study the list in the privacy of their own offices and to formulate their priorities for the Thursday meeting. They may also think of other factors that may be added to the list.

The Follow-up Session

While the brainstorming session is relatively short, the Thursday follow-up will take an entire morning. The meeting begins with a review of the goals of the experiment. A goal is the bottom-line result that we hope to attain from the experimentation. (e.g., "Make the thing work").

The remainder of the first hour of this session is devoted to the rather easy aspects of this meeting. Any additions from the private time the participants had on Wednesday are placed on the list. The absolute "off the wall," ridiculous ideas are purged from the list and any redundant ideas are consolidated.

The next part of the follow-up session is very important. This is the point when we actually select the factors that will become a part of the experimentation. The process is one of elevation rather than elimination. This positive approach is taken, since we will not be able to include all of the factors in a single experiment. If we eliminate factors, we may never come back to those factors in subsequent efforts. Various voting and ranking methods may be used to facilitate this process.

Once the finalists have been selected, it is important to identify the levels to be used in the experiment. If such information is not available, it may be necessary to try some levels out in preliminary testing before the main event of the experiment begins. Remember, the levels must function with all of the other factors at the ranges of levels over which they are set. Don't design an experiment that has runs that are impossible.

The final activity will help in the selection of the experimental configuration. If the team is able to focus their prior knowledge and to determine where there may be likely interactions, then the experimental structure can be made smaller. To determine if a possible interaction exists, ask the following question: Will the influence of this factor be the same on the response if I should change another factor? If the answer is yes then there is no interaction. If the answer is that there will be a difference in how the factor influences the response depending on the level of another factor, then there is an interaction. Sometimes a plot of the anticipated behavior of two factors on the response helps clarify this concept.

Now we have all the engineering inputs that will be focused by the experimental structures that we have learned in this chapter. The entire process we have just reviewed is an integral and essential part of the experimental design effort.

5
Building the Quality In — The Parameter Design

The pivotal concept in the quality engineering by design methodology is the control of variation while still holding to the target value. The loss function drives us in this direction. To control a product's characteristics, we must understand what causes those characteristics to change, take charge of those change factors, and adjust them to obtain the desired level of output. To gain the understanding, we must either go to the bank of knowledge and make a withdrawal, or make an investment in new knowledge via experimentation. Since most new products are pushing the state of the art, it is often necessary to extend our knowledge with the experimental approach. We learned how to build efficient experimental structures in the last chapter. Now we will put these experimental structures to work.

As engineers, we understand that there is a functional relationship between the factors under our control and the responses we measure. We will discover how to compensate for variation induced by factors beyond our control (noise factors) by the proper choice of the levels of the factors under our design authority (controlling factors). Experimental design will be used to quantify our hypotheses and to find the settings (or levels) of each factor that will optimize the process under study. The experimental effort that leads to the proper choice of levels is called the *parameter design,* for we are able to find the levels of the parameters (or factors) that put us on target with low variation.

To show how the parameter design works, let us revisit the image-quality (IQ) experiment from Chapter 4. Recall that we were investigating three factors (density, raggedness, width) at two levels each. The experimental layout was as follows:

Table 5-1.

Density	Raggedness	Width
light	rough	narrow
dark	rough	narrow
light	sharp	narrow
dark	sharp	narrow
light	rough	wide
dark	rough	wide
light	sharp	wide
dark	sharp	wide

We observed in Chapter 4 that the above design is quite efficient in detecting trends in single-factor effects as well as interactions. However, this experiment only looks at the influence of these effects at the midpoint or average level of each factor. If we were to repeat the experiment again using different observers and different images, we would probably get a different set of IQ results. This would be due to two obvious sources of variation, the observer-to-observer variation, as well as the inability to make the images exactly the same from one set to another set. We could insulate the experiment from the set-to-set variation by using the same set of images for all observers. However, if in reality there is variation in the image-making ability of the copying device that produced the images, we would only be able to draw conclusions for that set of images and would not have a solid enough experiment from which to draw general conclusions about these factors. Therefore, instead of looking at only one set of images with only one observer, we usually make many sets of images and gather responses from many observers. This method of collecting random samples does two things, it ensures generality and projectability of the results and it also gives us the ability to see if the differences and trends that we observe are any different from the variation that is randomly sprinkled over the entire experiment. If the variability is greater than the differences and trends, then we would have to conclude that the differences observed could only be chance happenings and should not be considered real influences. If the differences observed exceed the variation in the random samples, we may conclude that these differences are illustrating real effects.

The above description of the method of random sampling is the traditional statistical approach to experimental design. This approach is linked to "classical" statistical analysis methods such as the t test and the analysis of variance (ANOVA). This method makes good, logical sense when we are trying to separate the information about trends in the average values from the variation around these trends. However, Taguchi is interested in understanding the variation as a response in itself, as well as the information about the trends due to the averages. The goal of such a Taguchi-approach analysis would be to find the optimal conditions of the factor settings that will satisfy the response from both a location and a variation criterion. One way of determining this optimal is to locate the conditions that will minimize loss.

In order to study the location trends, we have learned to make purposeful changes to factors that influence the location response by their setting. If we also need to study the variation trends, we must make purposeful changes to the factors that will influence the variation response. Therefore, the Taguchi approach to experimental design has two objectives: a location-seeking objective and a variation-seeking objective.

We will now look at the IQ experiment example using the Taguchi approach. The first part of the experimental design will be exactly the same as we used before. However, we will add another part to this design to allow the investigation of variation. To produce such variation, we could take each run of the experiment and repeat it much in the same way as the traditional experimental design would dictate. However, in doing so, we may or may not obtain a representative sample of the degree of variation associated with that run. Since this lack of predictability of the variation could lead to inconsistent results, we will utilize the more innovative approach that Taguchi proposes. This approach induces the variation in a systematic manner via a designed experiment.

Where does variation come from? We have already agreed that it would be impossible to replicate each run of the experiment exactly, due to the variation induced by our inability to make the images the same if we tried to do so. This is the variation caused by inner noise we discussed in Chapter 3. Table 5-2 shows some possible sources of inner noise associated with this IQ experiment.

In the production of the images that are to be judged for quality, the density factor is influenced by the amount of pressure exerted by the marking instrument. Therefore, to understand the influence of the variation caused by pressure we will purposely exert light and heavy pressures when making the images in our noise design. The ragged images will be varied in a similar fashion with a greater degree of raggedness purposely induced according to an experimental design configuration. The third factor, width, will be exaggerated by its tolerance. We will make a certain portion of the images wider (by 1.5 times) than others according to the statistical experimental design plan.

All of the above variations are exaggerated and are actually forced to take place in our experiment. The reason for the exaggeration is twofold. If the wide variation (or low cost tolerance, as Taguchi calls it) gives us a product that works consistently (has low loss), then we have a robust product that is less expensive. The second reason is related to the fact that we are experimenting to optimize the variation of the product. If we are unable to observe changes in the product's characteristics, then our efforts to understand variation have failed. By making big changes, we ensure that if the factor is influential, it will emerge, cause change, and help us understand the sources of variation. This is somewhat contrary to the usual statistical approach to experimentation, which strives to keep the varia-

Table 5-2.

Factor	Low Cost Variation
Density	Light & heavy pressure
Raggedness	More ragged & normal raggedness
Width	1.0 (nominal) & 1.5 times wider

tion low, so as not to overpower the information on trends due to the average settings of the factors. However, Taguchi is interested in more than just the trends due to the the averages. Understanding trends due to the variation is just as important to the effort to reduce loss, so we provide an opportunity to study these trends in variation by systematically inducing variation so that it may also be investigated, measured, and controlled.

Now that we understand the engineering, statistical, and economic reasons for the controllable sources of noise (inner noise) being purposely induced in our experiments, there is still another type of variation that we discussed in Chapter 3 that cannot be ignored if we are to have robust product or process designs. That type of variation is the outer noise, a type of noise that is beyond our control, such as the environment or materials. A possible outer noise in our IQ experiment is the lighting conditions under which the images are judged. In low light, the images are harder to see than in bright light. Yet, we may be forced to use images in all types of lighting conditions, so we want a "quality" image that looks good consistently all the time. We will cause both inner and outer noises to take place in our structured approach to variation engineering. Table 5-3 shows the experiments we will use to do the job.

To help understand the information in this table, we introduce some terminology to describe the elements of the parameter design. The first part of a parameter design shown to the left in Table 5-3 is called the *inner array* by Taguchi. A more accurately descriptive name for this part of the parameter design was given by Raghu Kackar (2) of Bell Labs. He calls it the *design matrix.* The word *design* does not refer to a statistical experimental design, but rather to the fact that the factors and levels included in this part of the effort are actually the potential conditions that we are seeking for our final engineering design configuration. So the word *design* means engineering design. We note that the relative magnitude of the changes in this design matrix is large. We may name this first element of the parameter design, the inner array or design matrix where *macro* size changes are made. When we look closely at this particular design configuration for the factors of density, raggedness, and width, we observe that it is no different than the experiment we put together in Chapter 4 to study these factors. What distinguishes this experimental system of Taguchi's from the classical experimental methods is the variation-inducing *outer arrays* (as Taguchi calls them) or *noise matrices* (as Kackar has named them). This is the aspect of experimentation that sets Taguchi's approach apart from all others. Each run of the design matrix triggers a set

Table 5-3.

DENSITY	RAGGEDNESS	WIDTH
Light	Rough	Narrow
Dark	Rough	Narrow
Light	Sharp	Narrow
Dark	Sharp	Narrow
Light	Rough	Wide
Dark	Rough	Wide
Light	Sharp	Wide
Dark	Sharp	Wide

OUTER ARRAYS

NOISE MATRICES

micro CHANGES PLUS

OUTER NOISE FACTOR(S)

Pressure	Δ Roughness	Δ Width	LIGHT	Response Image Quality
Light	More	1.0	Dark	
Heavy	More	1.0	Bright	
Light	Nominal	1.0	Bright	
Heavy	Nominal	1.0	Dark	
Light	More	1.5	Bright	
Heavy	More	1.5	Dark	
Light	Nominal	1.5	Dark	
Heavy	Nominal	1.5	Bright	

Pressure	Δ Roughness	Δ Width	LIGHT	Response Image Quality
Light	More	1.0	Dark	
Heavy	More	1.0	Bright	
Light	Nominal	1.0	Bright	
Heavy	Nominal	1.0	Dark	
Light	More	1.5	Bright	
Heavy	More	1.5	Dark	
Light	Nominal	1.5	Dark	
Heavy	Nominal	1.5	Bright	

Pressure	Δ Roughness	Δ Width	LIGHT	Response Image Quality
Light	More	1.0	Dark	
Heavy	More	1.0	Bright	
Light	Nominal	1.0	Bright	
Heavy	Nominal	1.0	Dark	
Light	More	1.5	Bright	
Heavy	More	1.5	Dark	
Light	Nominal	1.5	Dark	
Heavy	Nominal	1.5	Bright	

Pressure	Δ Roughness	Δ Width	LIGHT	Response Image Quality
Light	More	1.0	Dark	
Heavy	More	1.0	Bright	
Light	Nominal	1.0	Bright	
Heavy	Nominal	1.0	Dark	
Light	More	1.5	Bright	
Heavy	More	1.5	Dark	
Light	Nominal	1.5	Dark	
Heavy	Nominal	1.5	Bright	

INNER ARRAY
DESIGN MATRIX
MACRO CHANGES

				Response
Pressure	Δ Roughness	Δ Width	LIGHT	Image Quality
Light	More	1.0	Dark	
Heavy	More	1.0	Bright	
Light	Nominal	1.0	Bright	
Heavy	Nominal	1.0	Dark	
Light	More	1.5	Bright	
Heavy	More	1.5	Dark	
Light	Nominal	1.5	Dark	
Heavy	Nominal	1.5	Bright	

				Response
Pressure	Δ Roughness	Δ Width	LIGHT	Image Quality
Light	More	1.0	Dark	
Heavy	More	1.0	Bright	
Light	Nominal	1.0	Bright	
Heavy	Nominal	1.0	Dark	
Light	More	1.5	Bright	
Heavy	More	1.5	Dark	
Light	Nominal	1.5	Dark	
Heavy	Nominal	1.5	Bright	

				Response
Pressure	Δ Roughness	Δ Width	LIGHT	Image Quality
Light	More	1.0	Dark	
Heavy	More	1.0	Bright	
Light	Nominal	1.0	Bright	
Heavy	Nominal	1.0	Dark	
Light	More	1.5	Bright	
Heavy	More	1.5	Dark	
Light	Nominal	1.5	Dark	
Heavy	Nominal	1.5	Bright	

				Response
Pressure	Δ Roughness	Δ Width	LIGHT	Image Quality
Light	More	1.0	Dark	
Heavy	More	1.0	Bright	
Light	Nominal	1.0	Bright	
Heavy	Nominal	1.0	Dark	
Light	More	1.5	Bright	
Heavy	More	1.5	Dark	
Light	Nominal	1.5	Dark	
Heavy	Nominal	1.5	Bright	

TABLE 5-4 A .

Pressure	Change in Roughness	Change in Width	Room Lighting	Image Quality
light	more	1.0	Dark	3
heavy	more	1.0	Bright	3
light	nominal	1.0	Bright	5
heavy	nominal	1.0	Dark	7
light	more	1.5	Bright	5
heavy	more	1.5	Dark	5
light	nominal	1.5	Dark	5
heavy	nominal	1.5	Bright	4

of runs in a noise matrix. The noise matrix contains tolerance-sized changes that are micro in scope when compared to the size of the changes in the design matrix. However, the noise matrix can contain more than just tolerance changes. It often (depending on the engineering needs) will contain outer noise factor changes in addition to the tolerance changes. As we stated before, by inducing variation from both inner sources (tolerance) and outer sources, we are able to find functional relationships between the design conditions and variation. With this knowledge of the functional dependence of variation, we are able to control the variation by choosing the best design levels of each factor. Yet, we must maintain a balance between the location (average output) of our response and its variation. If we were to optimize only on variation, or only on location, we would not have a total quality product or process. Therefore, we must utilize a measure that looks at both location and variation simultaneously. We know of such a quality metric. It is the loss function.

In the case of the IQ experiment, we are striving for higher quality (a bigger number) with low variation. The loss function will be one sided in this case. We will compute the k value for the loss and then compute an expected loss for each run of the design matrix.

We will measure IQ on a 1–10 scale, with 1 being poor and 10 being designated as good. In Appendix 5-1 of this chapter we will explain the details of the data-gathering procedure that produced the results shown in Table 5-4. Now it is important to look at these results and let the experiment help us understand how to make a robust imaging system. Such a system will produce low average (or expected) loss. We have eight poten-

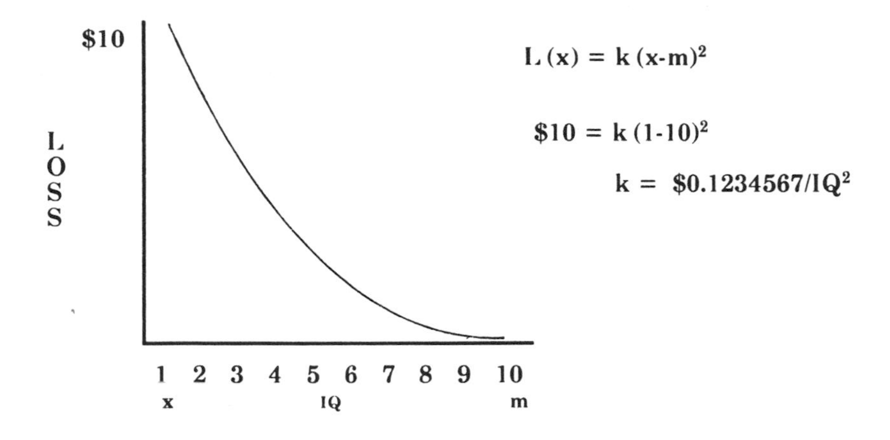

Figure 5-1.

tial systems to choose from our design matrix. Within each of the eight, there are eight variations around the design points. These variations include both tolerance changes and the outer noise variation, the light level. The design point from the design matrix will determine the average set point for the output (or response), while the small changes around this set point will produce the variation we wish to minimize. If we just scan through the 64 observations, we can readily observe that the last combination of conditions produces a set of responses that are quite high and exhibit very low variation. If we merely look at the best run of an experiment, we miss the understanding that the structured method brings to us. Therefore, we will analyze the entire experiment to see the direction toward our goal of high quality.

We will begin this analysis by summarizing the results of each noise matrix. We have computed the average and the standard deviation, which are shown in Table 5-4. However, to optimize for both of these at the same time is a somewhat tedious task, since it involves two separate analyses and in many cases a balancing or trade-off to find the best conditions to satisfy these two measures of quality simultaneously. The loss function combines the location measure (average) and the variation measure (standard deviation [SD]) in a single metric. When we minimize this loss, we will automatically make the trade-off between location and variation.

Table 5-4.

OUTER ARRAYS
NOISE MATRICES
micro CHANGES PLUS
OUTER NOISE FACTOR(S)

DENSITY	RAGGEDNESS	WIDTH
Light	Rough	Narrow
Dark	Rough	Narrow
Light	Sharp	Narrow
Dark	Sharp	Narrow
Light	Rough	Wide
Dark	Rough	Wide
Light	Sharp	Wide
Dark	Sharp	Wide

(1)

	Pressure	Δ Roughness	Δ Width	LIGHT	Response Image Quality
1	Light	More	1.0	Dark	3
2	Heavy	More	1.0	Bright	3
3	Light	Nominal	1.0	Bright	5
4	Heavy	Nominal	1.0	Dark	7
5	Light	More	1.5	Bright	5
6	Heavy	More	1.5	Dark	5
7	Light	Nominal	1.5	Dark	5
8	Heavy	Nominal	1.5	Bright	4

$\bar{\bar{X}} = 4.63$, $s = 1.30$, Lo$$ = 3.75, $S/N_B = 12.4$

a

	Pressure	Δ Roughness	Δ Width	LIGHT	Response Image Quality
1	Light	More	1.0	Dark	3
2	Heavy	More	1.0	Bright	4
3	Light	Nominal	1.0	Bright	7
4	Heavy	Nominal	1.0	Dark	4
5	Light	More	1.5	Bright	3
6	Heavy	More	1.5	Dark	2
7	Light	Nominal	1.5	Dark	4
8	Heavy	Nominal	1.5	Bright	3

$\bar{\bar{X}} = 3.75$, $s = 1.49$, Lo$$ = 5.06, $S/N_B = 10.0$

b

	Pressure	Δ Roughness	Δ Width	LIGHT	Response Image Quality
1	Light	More	1.0	Dark	3
2	Heavy	More	1.0	Bright	9
3	Light	Nominal	1.0	Bright	6
4	Heavy	Nominal	1.0	Dark	9
5	Light	More	1.5	Bright	3
6	Heavy	More	1.5	Dark	10
7	Light	Nominal	1.5	Dark	3
8	Heavy	Nominal	1.5	Bright	9

$\bar{\bar{X}} = 6.25$, $s = 2.96$, Lo$$ = 2.69, $S/N_B = 12.8$

ab

	Pressure	Δ Roughness	Δ Width	LIGHT	Response Image Quality
1	Light	More	1.0	Dark	8
2	Heavy	More	1.0	Bright	9
3	Light	Nominal	1.0	Bright	9
4	Heavy	Nominal	1.0	Dark	10
5	Light	More	1.5	Bright	9
6	Heavy	More	1.5	Dark	8
7	Light	Nominal	1.5	Dark	6
8	Heavy	Nominal	1.5	Bright	8

$\bar{\bar{X}} = 8.38$, $s = 1.19$, Lo$$ = 0.48, $S/N_B = 18.2$

INNER ARRAY

DESIGN MATRIX

MACRO CHANGES

c

	Pressure	Δ Roughness	Δ Width	LIGHT	Response: Image Quality
1	Light	More	1.0	Dark	5
2	Heavy	More	1.0	Bright	1
3	Light	Nominal	1.0	Bright	6
4	Heavy	Nominal	1.0	Dark	4
5	Light	More	1.5	Bright	3
6	Heavy	More	1.5	Dark	1
7	Light	Nominal	1.5	Dark	7
8	Heavy	Nominal	1.5	Bright	4

$\overline{X} = 3.88$
$s = 2.17$
Lo$$ = 5.14
$S/N_B = 5.4$

ac

	Pressure	Δ Roughness	Δ Width	LIGHT	Response: Image Quality
1	Light	More	1.0	Dark	1
2	Heavy	More	1.0	Bright	1
3	Light	Nominal	1.0	Bright	4
4	Heavy	Nominal	1.0	Dark	6
5	Light	More	1.5	Bright	2
6	Heavy	More	1.5	Dark	1
7	Light	Nominal	1.5	Dark	4
8	Heavy	Nominal	1.5	Bright	4

$\overline{X} = 2.75$
$s = 1.67$
Lo$$ = 6.79
$S/N_B = 3.6$

bc

	Pressure	Δ Roughness	Δ Width	LIGHT	Response: Image Quality
1	Light	More	1.0	Dark	4
2	Heavy	More	1.0	Bright	9
3	Light	Nominal	1.0	Bright	6
4	Heavy	Nominal	1.0	Dark	9
5	Light	More	1.5	Bright	5
6	Heavy	More	1.5	Dark	9
7	Light	Nominal	1.5	Dark	7
8	Heavy	Nominal	1.5	Bright	10

$\overline{X} = 7.38$
$s = 2.20$
Lo$$ = 1.37
$S/N_B = 16.1$

abc

	Pressure	Δ Roughness	Δ Width	LIGHT	Response: Image Quality
1	Light	More	1.0	Dark	9
2	Heavy	More	1.0	Bright	10
3	Light	Nominal	1.0	Bright	10
4	Heavy	Nominal	1.0	Dark	10
5	Light	More	1.5	Bright	10
6	Heavy	More	1.5	Dark	10
7	Light	Nominal	1.5	Dark	9
8	Heavy	Nominal	1.5	Bright	10

$\overline{X} = 9.75$
$s = 0.46$
Lo$$ = 0.03
$S/N_B = 19.8$

Let's step through a loss calculation for the first set of conditions of the inner array. This is a nominal setting of light, rough, and narrow. The noise matrix shown in Table 5-4A produces variation around this nominal setting. The loss for the first observation of the noise matrix where the IQ was reported at 3 is calculated as follows:

$$\text{Loss} = k(y-m)^2.$$

$$\text{Loss} = .1234567(3-10)^2 = \$6.05.$$

We compute the loss for each of the other seven observations. To obtain the average loss, we simply take the sum of these eight losses and divide by 8 to produce an average loss of $3.75. We would of course like our loss to be zero, so this is not a very desirable outcome. However, we must look at this outcome along with the others to find the trends for the best design of this imaging system. To do so, we will collapse the entire parameter design back down to the design matrix stage, and since we believe that the quality (as measured by the loss) is a function of the factors we have changed (we would not have run the experiment otherwise!), we will look for the trends in the loss as we change from light to dark, from rough to smooth, and from narrow to wide.

Table 5-5 shows these results. The analysis is a simple contrast between the two conditions. However, there does not seem to be much change for the density or line-width factors. We need to look more deeply into the analysis and to investigate the interactions between the factors.

Table 5-5.

Summary of Losses for Design Matrix.

Density	Raggedness	Width	Lo$$
light	rough	narrow	$3.75
dark	rough	narrow	$5.06
light	sharp	narrow	$2.69
dark	sharp	narrow	$0.48
light	rough	wide	$5.14
dark	rough	wide	$6.79
light	sharp	wide	$1.37
dark	sharp	wide	$0.03

Contrast analysis of loss data from Table 5-5:

Density

Average loss at:

light	$3.24
dark	$3.09

Raggedness

Average loss at:

rough	$5.19
sharp	$1.14

Width

Average loss at:

narrow	$3.00
wide	$3.33

As Figures 5-2 and 5-3 indicate, there is a large difference between the IQ of the rough and sharp images. However, the density factor relationship depends on the raggedness factor. As the image goes from light to dark, the quality either decreases (for a rough image) or increases (for a sharp image). Since sharp images are judged better than rough images, this interaction tells us to produce dark images as long as we can keep

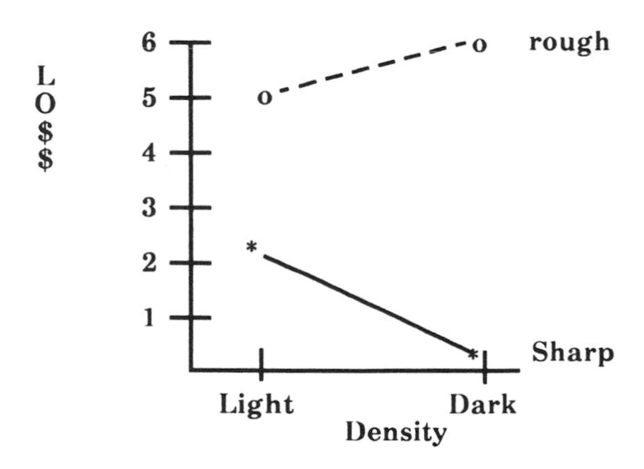

Figure 5-2.

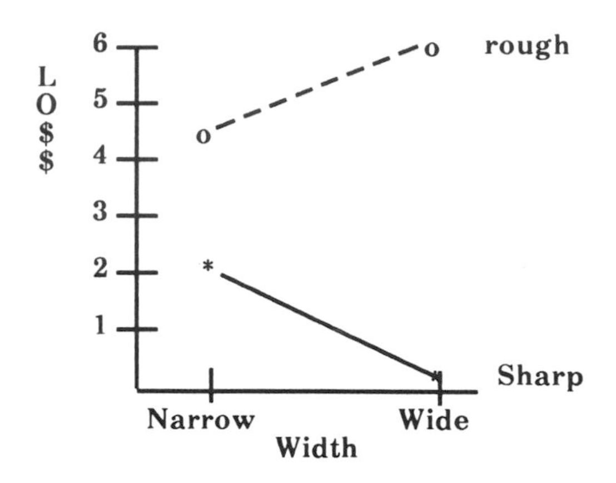

Figure 5-3.

them sharp. Figure 5-3 has a similar message concerning the width. As the image goes from narrow to wide, the quality either decreases (for a rough image) or increases (for a sharp image). Again, since sharp images are judged better than rough images, the interaction tells us to produce wide images as long as they are smooth. Remember, in all these decisions, we are calling the best (highest quality) conditions those that produce the lowest loss.

The final conclusions are to produce an image that is sharp, dark, and wide. This happens to be the set of conditions that appear in the last run of the design matrix. We could have simply looked at this run and picked it without the analysis we have just completed; however, the insight into the reasons for our decision are often important to ensure the longevity of our decisions. This desire to know the "mechanism" behind the optimal conditions is a manifestation of the scientist in every engineer. We will also see that in more complex experiments, involving fractional factorials, the trend analysis is the only way to come to the optimal conditions, since the best set of conditions often do not appear in the fractional design matrix.

One final observation must be made before we leave this example of the parameter design. The observer who made the IQ judgments was asked to write down his criteria for judging the images. He wrote as follows:

1–5 any line with smudge	5–7 light sharp
5–10 any sharp one	7–10 dark sharp

When this observer was told the results, he was amazed that the experiment was able to uncover his criteria exactly! In general this shows that statistical experimental design is a most powerful investigation tool. Complex, human decisions are some of the hardest to understand. These methods have certainly done the job of helping us to understand the thought process that made the judgments! If design of experiments methods are able to help understand a complex mechanism like the human thought process, they certainly should be able to help with other complex processes such as electronic devices, automotive designs, mechanical engineering problems, as well as chemical engineering efforts.

Experimental design has been used successfully in all of these areas over the past 50 years. It is through the efforts of Dr. Taguchi that the methods have become more engineering oriented and less dependent on the presence of a full-time statistician. As we move to more examples we will see how easy it is to apply statistical experimental design to such problems. Also, the full statistical analysis for each example is shown in Appendix 5-2. These analyses point out the presence of the interactions we plotted and that lead to the conclusions we drew.

In the last example, we used a two-level experimental structure for both the design matrix and the noise matrices. We also ran an experiment based on gathering "live" data. In this next example we will expand our design to three levels. In this way, if there are any curved relationships, we will be able to discern and make use of them. The second difference in this new example will be in the way the "data" are accumulated. Instead of making actual physical changes, we will utilize a known, physical-based relationship. In this way, we will be able to enhance a product design without actually testing that product. All the "testing" will take place on paper. This example illustrates an important point about the Taguchi optimization process:

THESE METHODS MAY BE USED FOR BOTH PHYSICAL EXPERIMENTATION AS WELL AS MATHEMATICAL MODEL OPTIMIZATION.

The exciting part of this important point is the fact that the *same* technique (design matrix/noise matrices) is used for both experimentation and model optimization. The process that we will optimize is a simple electronic circuit. The impedance (Z) of an alternating current circuit is given as follows:

$$Z = \sqrt{R^2 + \left(\frac{1}{2\pi(f \cdot C)}\right)^2}, \tag{5-1}$$

where R is the resistance of the circuit measured in ohms, $\pi = 3.14159$, f is the frequency measured in hertz, and C is the capacitance measured in farads.

The specifications for this circuit call for an impedance of 550 ohms with minimum loss due to variation and being off target. We will also want to build the lowest cost circuit with tolerances on the components that are inexpensive. We choose the following "low-cost" tolerances for the three components of the circuit.

Table 5-6.

Component	Low-Cost Tolerance
Frequency	10%
Capacitance	20%
Resistance	1%

Through some preliminary calculations, we determine the starting values for the frequency and capacitance. The value of the resistor to reach the 550-ohm requirement will be dependent on the choices of f and C for each design configuration. We will make rather large changes (each level will be double the previous level) in this optimization procedure. This is possible, since there will be no real physical experimentation going on. Also, we must perturb the system if we ever expect it to "talk" to us.

The starting values and their ranges are as follows:

Frequency	60	120	240 hertz (Hz)
Capacitance	5	10	20 microfarads (mfd)

Since there is the possibility of a curvilinear effect, we will utilize a three-level experimental structure to find the best design point for the frequency and capacitance. Because there are only two factors, we may use them L9 design without any confounding concerns. The design matrix is as follows.

Table 5-7.

Run #	Frequency	Capacitance
1	60 Hz	5 mfd
2	60 Hz	10 mfd
3	60 Hz	20 mfd
4	120 Hz	5 mfd
5	120 Hz	10 mfd
6	120 Hz	20 mfd
7	240 Hz	5 mfd
8	240 Hz	10 mfd
9	240 Hz	20 mfd

As in the previous example, each run of the design matrix will trigger a noise matrix in order to generate the expected loss due to location shifting and variation. We will show the runs for the first run of the design matrix, but only the results for the remaining runs. In this case we will use another L9 structure for the noise matrices. However, since we will need to include three factors (frequency, capacitance, and now the resistance), we will use a fractionally factorial version of L9. The confounding due to this fraction is considerable, but is of no consequence in the noise matrix application, since we do not intend to analyze the results for trends. We are only interested in causing systematic changes to the factors according to their low cost variation. The use of a three-level design instead of a two-level design gives us the ability to include the midpoint or nominal level in the noise matrices. If resources are constrained, the use of a two-level design may be considered. In this example, the resources are not constrained because the "experiment" is done via calculation. The first noise matrix is Table 5-8.

We will need a loss function to evaluate our experimental results. The loss may be determined if the circuit is discarded. If the impedance drops below 542.5 or exceeds 557.5, it is considered nonfunctional and will be discarded. The cost of the components in this circuit is as follows:

Frequency generator: \$20.00
Capacitor: \$ 1.00
Resistor: \$ .12
Total: \$21.12

The k for this loss (assuming a quadratic loss function) is:

$$\text{Loss} = k(y-m)^2.$$

$$\$21.12 = k(542.5 - 550)^2.$$

$$k = \$21.12/(-7.5)^2.$$

$$k = \$.375.$$

We will now take the responses (Z values) and compute the loss for each, add up these losses, and divide by 9 (the number of observations in the noise matrix) to compute the estimate of the expected loss.

The loss for this first design is quite high because the Z values deviate widely from the aim value of 550 ohms, even though the average impedance (568.37) is close to the aim. This once again shows how important a part the variation (the SD is 104.85!) plays in the quality of the product. We complete the remaining loss calculations for the other eight runs of the design matrix. Table 5-10 shows the loss, average, and SD for each run of the design matrix.

Figure 5-4 is the plot of the results from Table 5-10, and we can immediately see that as frequency and capacitance are increased, the loss decreases. This decrease is not at the same rate for each frequency or capacitor. There is an interaction as well as a curved effect taking place with this function. (Of course we knew that by simply inspecting the function).

Table 5-8.

Frequency	Capacitance	Resistance	Impedance
54	4	143.64	750.7
54	5	145.09	607.1
54	6	146.54	512.6
60	4	145.09	678.8
60	5	146.54	550.4
60	6	143.64	464.8
66	4	146.54	620.4
66	5	143.64	503.2
66	6	145.09	427.3

Table 5-9.

Frequency	Capacitance	Resistance	Impedance(Z)	Loss
54	4	143.64	750.7	$15105.11
54	5	145.09	607.1	1220.50
54	6	146.54	512.6	524.26
60	4	145.09	678.8	6223.94
60	5	146.54	550.4	.05
60	6	143.64	464.8	2719.59
66	4	146.54	620.4	1859.09
66	5	143.64	503.2	820.64
66	6	145.09	427.3	5646.65

Average Impedance (Z)= 568.37; s = 104.85

ESTIMATE OF EXPECTED LOSS= $3791.09

Table 5-10.

Run #	Frequency	Capacitance	Estimate of Expected Loss	Average	Std. Dev.
1	60 Hz	5 mfd	$3791.09	568.3	104.8
2	60 Hz	10 mfd	265.41	556.2	27.4
3	60 Hz	20 mfd	22.11	551.6	7.9
4	120 Hz	5 mfd	265.41	556.3	27.4
5	120 Hz	10 mfd	22.11	551.7	7.9
6	120 Hz	20 mfd	7.90	550.4	4.8
7	240 Hz	5 mfd	22.11	551.7	7.9
8	240 Hz	10 mfd	7.90	550.4	4.8
9	240 Hz	20 mfd	7.44	550.1	4.7

We will expand the scale of the plot in Figure 5.5 to examine the relationship in the area of interest where the loss is low. This will allow insights into the trends that will lead us to the best settings for this circuit.

We can see that there is a rather rapid change in the loss, falling steeply from the low frequencies and low capacitances. However, the degree of change between the middle and high frequency and the middle and high capacitance is far less than the change between the low and middle levels of both of these factors. Now we will apply some economic en-

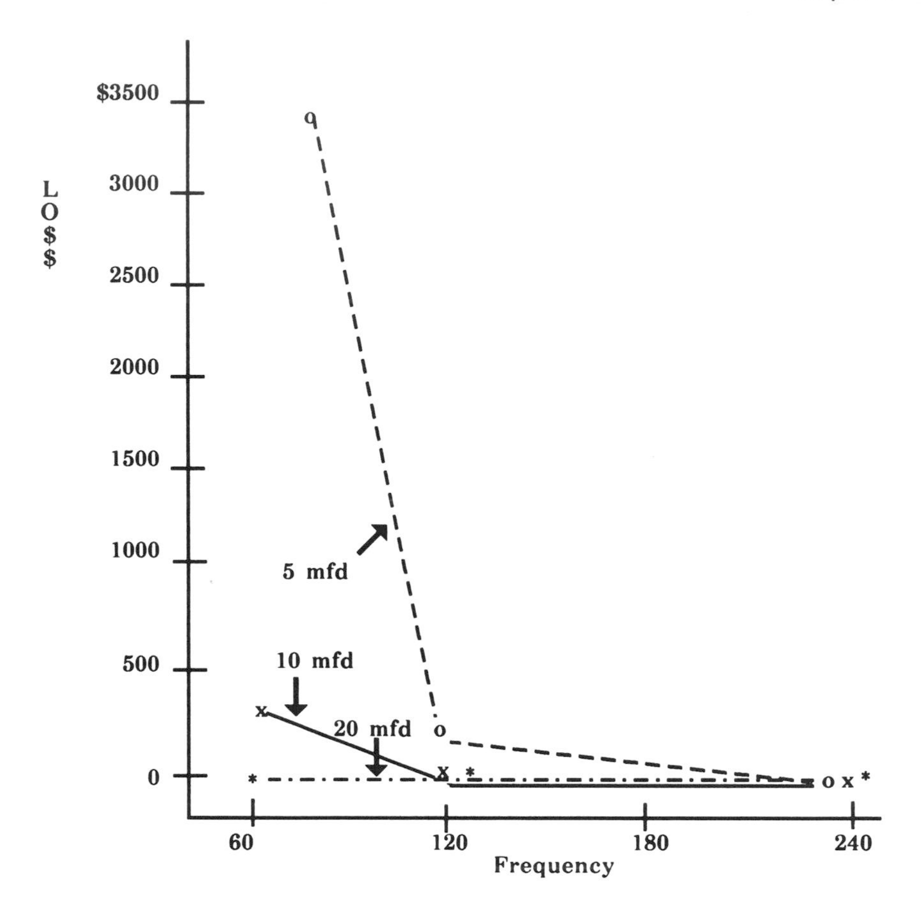

Figure 5-4.

gineering to this problem. An electrolytic capacitor gets bigger as the capacitance increases. The cost also increases. Let's say that a 10-mfd capacitor costs $1 and a 20-mfd costs $2. The difference in the loss between a 10-mfd and a 20-mfd capacitor when the frequency is set at 240 Hz is only 46 cents. Therefore by going to the lower cost capacitor we have a net saving of 54 cents per unit. However, if the frequency is set at 120 Hz, the 10-mfd capacitor has a loss of $22.11 vs. the loss of $7.90 for the 20-mfd one. This is of course the result of the strong interaction between the frequency and the capacitance and would drive the design of this circuit to

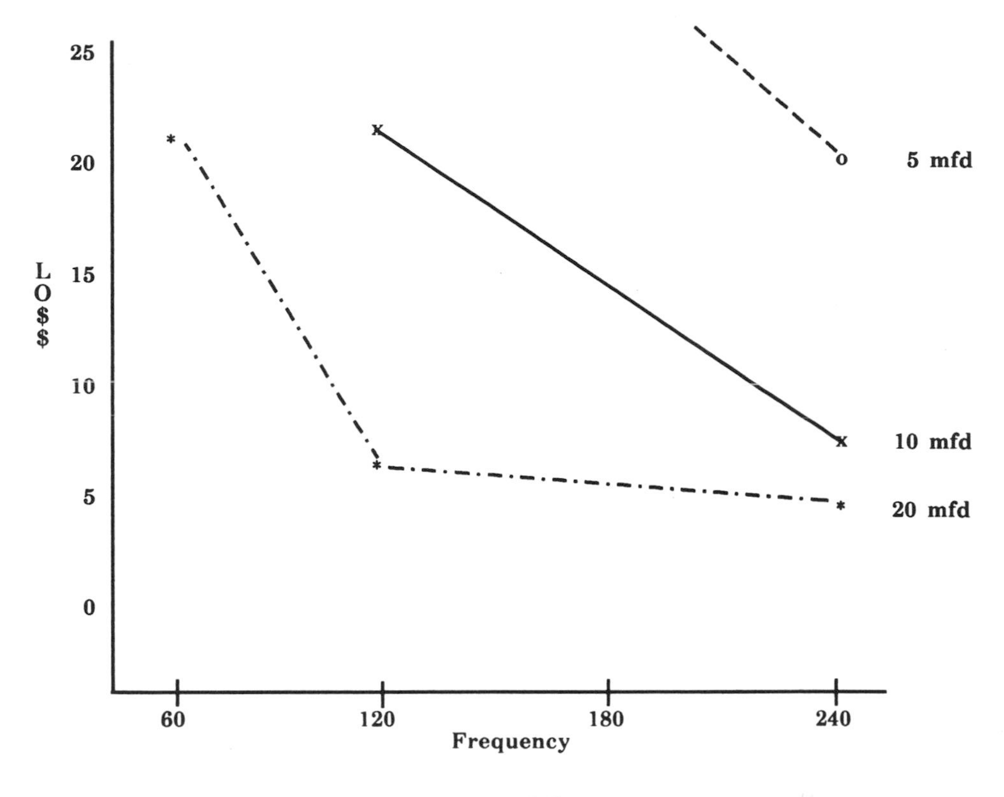

Figure 5-5.

the higher frequency. Since the frequency level is *not* cost dependent, we are able to achieve a low-loss (i.e., high-quality) design at a low cost.

Has all of this happened by some exercise of a magical statistical method? Certainly not! What we have observed is the physics of the problem obeying fundamental science. The statistics has simply allowed us to see this science take place. The frequency and the capacitor are the major contributors to the variation of the final impedance, since the tolerances on these components are wide. Therefore, the parameter design drives the engineering design to the point where these two components have a minimum contribution to the overall impedance. The conditions that lead to the low loss configuration are as follows:

Frequency = 240 Hz ╲
╱ ---- 0.18% of impedance,
Capacitance = 20 mfd

which leads to a resistance of 549, which is 99.82% of impedance.

Since the resistor has a 1% tolerance, the engineering design has gravitated to making this component the major contributor to the overall impedance. We have masked the effect of the other two (wide-tolerance) components to our advantage. Now, of course, if this masking has caused another desirable quality characteristic to be diminished, we must make the trade-off between its loss and the loss due to the variation in the impedance characteristic.

ANOTHER APPROACH TO ANALYSIS

While the method outlined in the preceding two examples utilizes the average loss as the figure of merit, there is another figure of merit that is related to the loss function and may be used to lead us to the correct quality engineering decisions. This figure of merit is the signal to noise (S/N) that we studied in Chapter 2. The reason for using the S/N rather than the loss stems from the problem of obtaining a loss function. In many cases, we do not know how much loss will be sustained when a quality characteristic (such as the IQ or the impedance) draws away from its optimal level. It would be possible to build a "relative" loss function to allow comparisons between the various design points, but such a relative loss would not allow us to make a meaningful cost/quality comparison, as we just did with the impedance circuit. While the loss function is the straightforward approach to the analysis of quality engineering by design problems, the use of the S/N is practiced by Taguchi and many of his followers. We shall therefore illustrate the S/N in action and show how it, in conjunction with the average level of the response, parallels the loss metric conclusions. We will take each of the parameter design examples and apply the appropriate S/N and then compare these results with those obtained using the loss function. While we minimized the loss, we will of course maximize the S/N.

Table 5-11 is exactly the same form as Table 5-5, except that in place of the loss response we have listed the type B S/N as well as the average value of the IQ. We will look at the same type of trend analysis for these responses as we did with the loss.

Table 5-11.

Density	Raggedness	Width	Average IQ	S/N(B)
light	rough	narrow	4.6	12.4
dark	rough	narrow	3.8	10.0
light	sharp	narrow	6.3	12.8
dark	sharp	narrow	8.4	18.2
light	rough	wide	3.9	5.4
dark	rough	wide	2.8	3.6
light	sharp	wide	7.4	16.1
dark	sharp	wide	9.8	19.8

CONTRAST ANALYSIS of IQ and S/N(B) from Table 5-11:

	For average IQ	For S/N(B)
Density		
Light	5.55	11.7
Dark	6.20	12.9
Raggedness		
Rough	3.78	7.9
Sharp	7.98	16.7
Width		
Narrow	5.78	13.4
Wide	5.98	11.2

As Figures 5-6 and 5-7 indicate, there is a large difference between the IQ of the rough and sharp images. However, the density factor relationship depends on the raggedness factor. As the image goes from light to dark, the quality either decreases (for a rough image) or increases (for a sharp image). Since sharp images are judged better than rough images, this interaction tells us to produce dark images as long as we can keep them sharp. Figure 5-7 has a similar message concerning the width. As the image goes from narrow to wide, the quality either decreases (for a rough image) or increases (for a sharp image). Again, since sharp images are

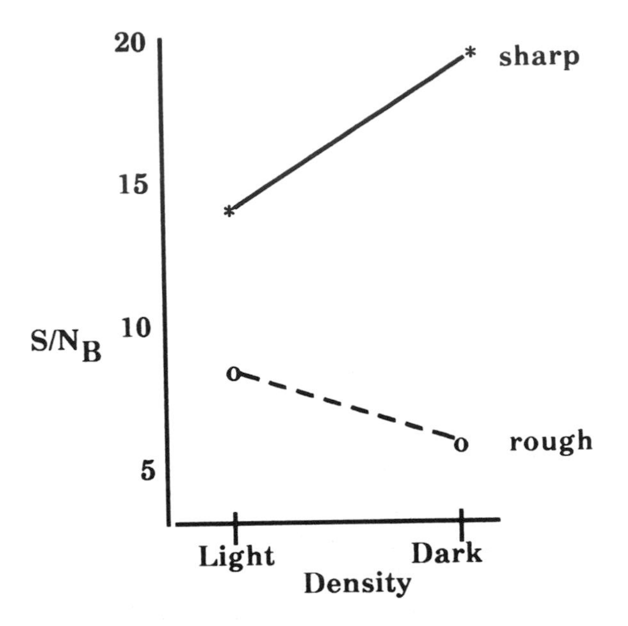

Figure 5-6.

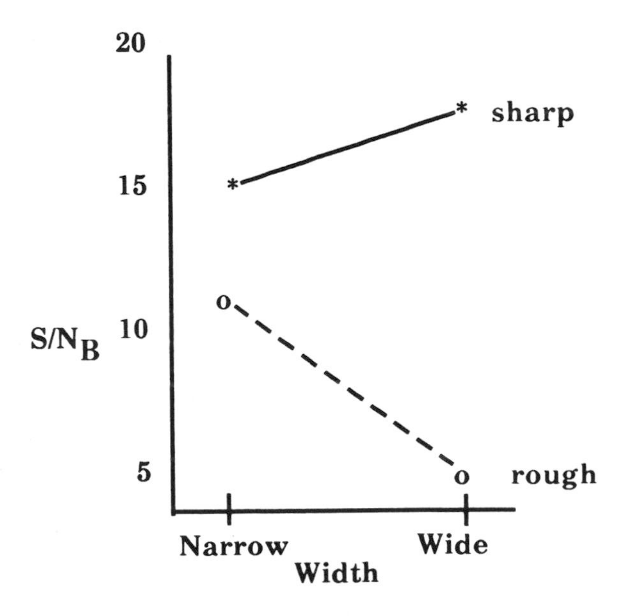

Figure 5-7.

judged better than rough images, the interaction tells us to produce wide images as long as they are smooth. Remember, in all these decisions we are calling the best (highest quality) conditions those that produce the highest S/N figure of merit rating.

The final conclusion leads us to recommend an image that is sharp, dark, and wide. This happens to be the set of conditions that appear in the last run of the design matrix and is exactly the same conclusion that we drew when we used the loss as the figure of merit. The fact that we obtained the same conclusion is not unexpected, since the S/N is merely a generalized transform of the loss function that we may use if an explicit loss function is not available. The disadvantage of using the S/N is the inability to make cost trade-offs, since we do not have a cost figure available. We will now look at the second example using the S/N as the figure of merit to optimize the design of the impedance circuit.

When we plot the results from Table 5-12 in Figure 5-8, we can immediately see that as frequency and capacitance are increased, the S/N(T) increases. This increase is not at the same rate for each frequency or capacitor. There is an interaction as well as a curved effect taking place with this function. This is the same interaction we observed when we used the loss as the figure of merit and leads us to the same conclusion.

Before we leave this chapter on the parameter design, we should make one further comparison of the analysis using the loss function and the alternative utilizing the S/N figure of merit. The loss function showed a rather wide fluctuation (especially in the impedance problem) to the function, while the S/N had a tendency to exhibit a more compact response.

Table 5-12.

Run #	Frequency	Capacitance	Estimate of Expected Loss	Average	S/N_T
1	60 Hz	5 mfd	$3791.09	568.3	14.4
2	60 Hz	10 mfd	265.41	556.2	25.8
3	60 Hz	20 mfd	22.11	551.6	36.6
4	120 Hz	5 mfd	265.41	556.3	25.8
5	120 Hz	10 mfd	22.11	551.7	36.6
6	120 Hz	20 mfd	7.90	550.4	41.1
7	240 Hz	5 mfd	22.11	551.7	36.6
8	240 Hz	10 mfd	7.90	550.4	41.1
9	240 Hz	20 mfd	7.44	550.1	41.3

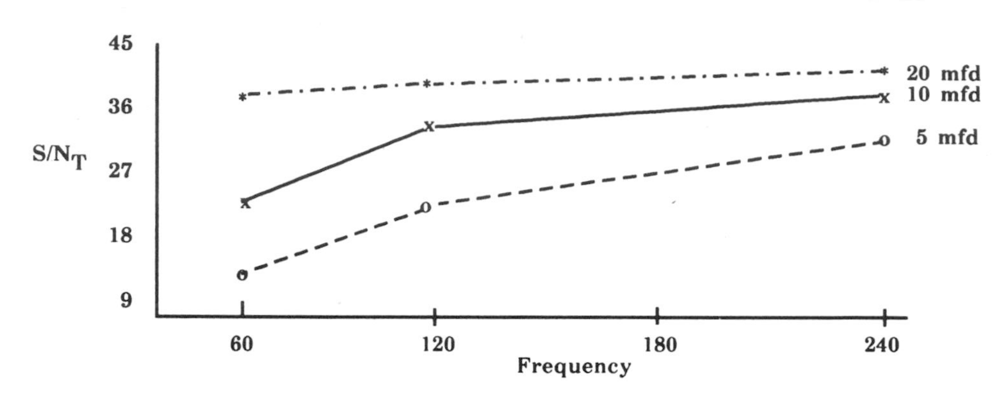

Figure 5-8.

This is, of course, due to the compression characteristic of the logarithmic function and helps in the plotting of the data. We did not need to expand the scale for the S/N plot in the impedance problem as we had to when the loss was used for this same problem. While not an overriding consideration, this ability to better communicate the results on a more conpact plot with the S/N is part of the reason for the use of the log in the S/N calculation.

As we have seen in the two examples, the parameter design is a very powerful method that gets us to the best design settings very rapidly and with a minimum of complicated analysis. All we had to do was plot the data and watch the trends. In many problems, though, there will be trends that are less explicit. This is where statistical analysis is necessary to separate the trend from other sources of unknown variation that can creep into our experiments. We certainly would not want to follow a trend that was untrue. Following such trends can be costly. We may decide to install equipment based on an experimental result that is wrong. The equipment may do nothing to improve the situation, or even cause a worse condition to take place. In the appendix, we present the statistical analysis that tells us if and where the most probable trends take place. It is important that similar analyses are used in each parameter design. This will prevent errors of interpretation from taking place and will allow the parameter design to lead us to the correct decisions.

REFERENCES

1. Taguchi, G. *Introduction to Quality Engineering.* Asian Productivity Organization, Tokyo, 1986.

2. Kackar, R. N. Off-line quality control, parameter design, and the Taguchi method, *J. Qual. Technol.* 17(4), 1985.

PROBLEMS FOR CHAPTER 5

1. Propose a parameter design for the following ink formulation used in an ink-jet computer printer. Write a statement of intent to be used in obtaining resources to fund this experiment. The goal of the effort is to reduce the blur in the image made by the ink-jet printer by reducing the number of free particles of pigment (carbon black) that surround the ink droplets. Do you think your request for resources will be approved?

Factors	Levels	Low-cost variation
Surface tension	20, 30, 40 dyn/cm^2	SD = 3 dyn/cm^2
Viscosity	5, 8, 11 centipoise	SD = 1 centipoise
Pigment particle size	2, 4, 6 μm	SD = .5 μm
Number of particles	10, 20, 30 per μl	SD = 1 particle
Printer drive power	.5, .75, 1.0 A	SD = .05 A
Storage	1, 13, 26 weeks	Outside noise

   Storage of ink before use is an outside noise factor that could range between 1 and 26 weeks.

2. The breakdown voltage of a semiconductor device has been shown to be a function of the following factors. Propose a parameter design to find the engineering design configuration to maximize breakdown voltage and to reduce the variation in this response variable.

Factors	Levels	Low-cost variation
Length	0, 1, 2 cm	±20%
Height	−2, 0, +2 cm	±20%
Width	0, 2, 4 cm	±20%
Temperature	160, 170, 180°C	±10%
Speed of rotation	1, 3, 5 rpm	±5%

Raw material purity	1, 3, 5 ppm	No control, but less than 5 ppm when in production
Moisture in atmosphere		No control during production and it will change from desert dryness (20% relative humidity [RH]) to 85% RH.

3. I would like to increase the number of miles per gallon (mpg) of gasoline in my automobile by improving my driving habits. Given the following possible factors that could relate to this opportunity, devise a parameter design that would accomplish this goal.

Factors	Possible levels	Low-cost variation
Speed	40–60 mph	5 mph
Fuel octane	87–92	1
Air conditioner	off-on-economy	none
Outside temperature	32–90°F	3°F
Outside humidity	20–90% RH	5% RH

Think of two more factors that could influence the mpg and include them in your design. How long do you think such an experiment would take? When would such an experiment be best planned to start during the year?

4. The charge voltage in an electrostatic copier depends on the position of the charging wire. The quality of the copy depends on the charge voltage. If the charge is too low, the image will be light. If the charge is too high, the image will be very black, but there will be an accompanying high level of background, which is a gray mist all over the copy. The mechanical engineer must position the bracket that holds the wire to impart the correct charge consistently. Devise a parameter design to accomplish this goal.

Factors	Working range	Low-cost variation
Bend angle of the bracket	87–91°	.5°

Thickness of metal	1.9–2.2 mm	.05 mm
Hole locations (from reference points)		
Unthreaded	14.5–16.0 mm	.05 mm
Threaded	14.0–16.0 mm	.05 mm
Torque on fasteners	90–110 in. oz.	2 in. oz.
Outside noise factors		
Relative humidity	30–70%	
Output current of power supply	20 μA	.5 μA

APPENDIX 5-1

METHOD FOR MEASURING IMAGE QUALITY

While there are many methods for putting a numerical value on the quality of an image ranging from very simple "go, no-go" judgments to complex measurements based on microdensitometry, we will utilize a customer-based measurement. The customer is shown the images that will be judged and is instructed to establish a scale from 1 to 10. This activity gives the observer a chance to see the range of IQ and also to gain experience from the practice run. The customer is told that the scale is a relative rating of this particular set of images. Therefore, he should not be making a comparison beyond this set. Once the customer has established the scale and the criterion for this scale, the experiment begins.

The set of images is randomized and presented one at a time to the customer. A relative rating value is given by the customer for each image and this is recorded as shown in Table 5-4. The values obtained are spread over a wide enough range to consider the response to be continuous enough to pursue statistical techniques based on interval methods.

APPENDIX 5-2

STATISTICAL ANALYSIS OF IMAGE-QUALITY PARAMETER DESIGN

The following is typical of the output of an analysis of means by using the variance method to determine the statistical significance of the factor on the response. Statistical significance simply means that the change induced by the factor is bigger than change induced by any other chance occuring variables. When the F ration exceeds a certain tabulated value, the probability of the factor influencing the response is at least the confidence level stated in the table.

Greg's Average Image-Quality Response

Total observation	Sum of square	Half effect	Measures	Df	Mean square	F ratio
4.6	276.1250	5.875	Average			
3.8	0.8450	0.325	A density	1	0.8450	1.690
6.3	35.2800	2.100	B raggedness	1	35.2800	70.560
8.4	5.1200	0.800	AB	1	5.1200	10.240
3.9	0.0800	0.100	C width	1	0.0800	0.160
2.8	0.0000	0.000	AC	1	0.0000	0.000
7.4	2.2050	0.525	BC	1	2.2050	4.410
9.8	0.0450	0.075	ABC	1	0.0450	0.090
Outside estimate of error					0.5000	

Plot points for A: Low 5.55, high 6.2.

Plot points for B: Low 3.775, high 7.975

Plot points for C: Low 5.775, high 5.975

Interaction table for the AB (Density · Roughness) Interaction

Level of Density	Level of Roughness	Point to plot
−	−	4.25
+	−	3.3
−	+	6.85
+	+	9.10

Interaction table for the BC (Roughness · Width) Interaction

Level of Roughness	Level of Width	Point to plot
−	−	4.2
+	−	7.35
−	+	3.35
+	+	8.60

Greg's S/N(B) Response

Total Observation	Sum of square	Half Effect	Measures	Df	Mean square	F ratio
12.4	1207.8610	12.288	Average			
10.0	3.0012	0.612	A density	1	3.0012	3.001
12.8	157.5313	4.438	B raggedness	1	157.5313	157.531
18.2	22.1112	1.662	AB	1	22.1112	22.111
5.4	9.0313	−1.063	C width	1	9.0313	9.031
3.6	0.1513	−0.138	AC	1	0.1513	0.151
16.1	41.8613	2.288	BC	1	41.8613	41.861
19.8	0.6613	−0.288	ABC	1	0.6613	0.661
Outside estimate of error					1.000	

Plot points for A; low 11.675, high 12.9.

Plot points for B: low 7.850, high 16.725.

Plot points for C: Low 13.35, high 11.225

Interaction table for the AB (Density · Raggedness) Interaction

Level of Density	Level of Raggedness	Point to plot
−	−	8.89
+	−	6.80
−	+	14.45
+	+	19

Interaction table for the BC (Raggedness · Width) Interaction

Level of Raggedness	Level of Width	Point to plot
−	−	11.2
+	−	15.5
−	+	4.5
+	+	17.95

Appendix Reference

Barker, T. B. *Quality by Experimental Design.* Marcel Dekker, New York, 1985, Chap. 13 and 14.

6
Tolerance Design

Taguchi has a planned attack on the sources of quality deterioration, called the *off-line quality control* method. The concept of this systematic approach to built-in quality is to design the product, process, or service properly before it ever gets into production. He begins this method with the *system design,* engineers the variation out of the system with the *parameter design,* and then, if necessary, utilizes the *tolerance design* to further reduce the variation to the required level.

Figure 6-1 shows the three stages of product/process/service design. We have devoted a considerable amount of this book to the second stage, parameter design, and very little to system design. In the practice of engineering in the United States, system design is done quite well. It is the product of fertile, inventive minds that thrive on breaking rules and trying innovative approaches to solving challenging design problems. However, creativity and discipline are often unable to peacefully coexist. This is one of the reasons that engineers in the United States are likely to "invent" a solution to a design problem rather than to apply the parameter design method. However, the Japanese engineer has a cultural background of rigid discipline and readily accepts the parameter design methods as ways

SYSTEM DESIGN	PARAMETER DESIGN	TOLERANCE DESIGN
Find the basic methods to accomplish a required output	Engineer to find the best settings that put the required output on target with low variation	Identify the quality sensitive components and apply tolerances to these components to meet the required level of variation on this output

Figure 6-1.

of improving the quality of the product or process. On the other hand, the culturally induced discipline of the Japanese stymies invention and creativity, although this is gradually changing, and there is evidence that the Japanese are producing more and more patents on basic concepts. Yet, Taguchi admits that his methods are not applicable to the system design.

The engineer in the United States still has the advantage during the system phase or invention process. Yet, this same engineer must then bring his invention to the market. He could do so by utilizing the parameter design as we have shown in the previous chapter, or he could continue to use the current practice. What is the current practice in the United States? In many cases, the engineer in the United States comes up with an invention, and then to make it work, he "gold plates" the invention with the most costly components and assembly methods. The antiquated idea that quality must be expensive permeates through such thinking. The U.S. engineer essentially goes straight from the system design into the tolerance design. In doing so, he misses the most important engineering activity, the parameter design. The parameter design is the part of the development process that the Japanese engineer does best and provides the advantage that has made Japan such a quality and productivity leader.

However, it is far more than the absence of parameter design that makes product development inefficient in the United States. It is the way the uniformed engineer applies tolerances that leads to even more costly products. A blanket set of very narrow tolerances are usually imposed on all the components. This is a very easy way to catch the quality-sensitive components (those that need tight tolerances), but it also makes the manufacturing cost higher for those components that do not need to be set so tightly. Good engineering procedure dictates that a sensitivity analysis be performed on a design to determine the relative contribution to the overall variation by each component. Then, a determination must be made to evaluate the cost of controlling the overall variation by taking control of these individual contributors to the variation. However, traditional sensitivity analysis is a costly engineering endeavor and is often only possible if a mathematical model of the process is available. This is where Taguchi's adaptation of statistical experimental design has made the required sensitivity analysis possible.

In our first example, we will utilize a mathematical model in order to illustrate how this adaptation of statistical experimental design accomplishes the objectives of a sensitivity analysis. We will extend the impedance example from the previous chapter by taking the best set of design conditions (the set with the lowest loss or highest signal to noise) [S/N]. This was determined to be the 240-Hz, 20-mfd, 549-ohm configuration. The standard deviation (SD) of the nine runs in the noise matrix was 4.72. The noise matrix is reproduced in Table 6-1 below.

If the customer who is purchasing this impedance circuit requires a level of variation that is half as large as the 4.72 ohms we have attained, we

Table 6-1.

Frequency	Capac.	Resistor	Impedance
216	16	543.5	545.4571
216	20	549.0	550.2344
216	24	554.5	555.3390
240	16	549.0	550.5619
240	20	554.5	555.4801
240	24	543.5	544.2116
264	16	554.5	555.7683
264	20	543.5	544.3450
264	24	549.0	549.5740

Average Impedance = 550.1
Standard Deviation = 4.72

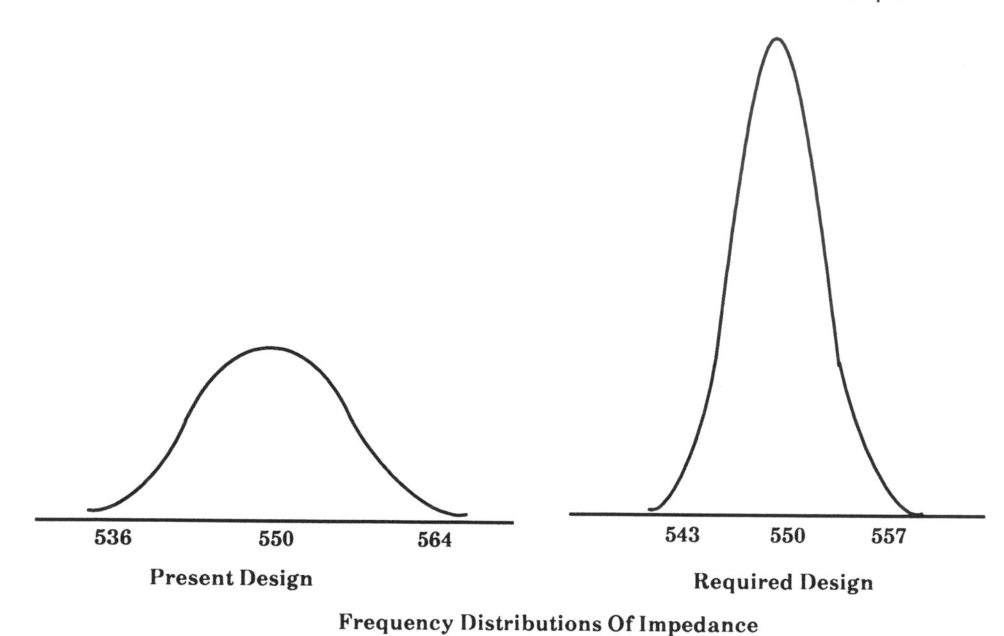

Figure 6-2.

would need to adjust our design to obtain a 2.36-ohm SD. Figure 6-2 shows the frequency distribution for the present design and the required distribution that will meet the customer's needs.

One method of obtaining the required tolerance is to simply measure the impedance values on each assembly and to deliver only those that meet the customer's requirements. While this method is frowned upon by most quality-control experts, it is commonly practiced in American industry. It is an expedient to "get the product out the door." The problem with such an approach is the expense added to the cost of doing business. There is the direct cost of paying the inspectors to sort the product, as well as the cost of paying the trash man to tote the rejects away. The number of outlet and surplus stores is a mute testimony to this attitude of sorting to different levels of quality. Sorting is not the long-term answer to the problem.

Disguised as a long-term solution to the sorting problem is the gold-plate method. In this approach, all components in the device are upgraded in quality by tightening their tolerances. In our example, the low

Table 6-2a.

Component	Variation	Cost
Frequency	+/-10%	$20.00
Capacitor	+/-20%	$ 1.00
Resistor	+/- 1%	$ .12

Current Design

Table 6-2b.

Component	Variation	Cost
Frequency	+/- 5%	$30.00
Capacitor	+/-10%	$ 2.00
Resistor	+/- .5%	$ .24

"Quality Design"

cost tolerances for the three components are found in Table 6-2a. The tighter tolerances that would satisfy the final product limits of variation are shown in Table 6-2b. A cost comparison from these tables shows that a total increase of $11.12 is necessary to upgrade to the levels that will achieve our goal.

While the "quality by dumping money at the problem" approach can achieve our goal and satisfy the customer, it is costly and wasteful. In this case, the extra cost of $11.12 could erase the profit on the product. Often the attitude is, "next time we will make the profit. Now we must hold market share no matter what the cost." However, there needs to be little if any added cost if the proper sensitivity analysis is performed. This simple example has been chosen because the sensitivity analysis may be done quite easily by just looking at the data in Table 6-1. The average impedance for this design is 550.1. The average resistance, however, is 549. The resistor is the major (99.8%) contributor to the impedance. Therefore, it must be the major contributor to the variation in the impedance. The *sensitivity analysis* requires nothing more than simple logic to accomplish its goal. We may achieve the low variation that the customer needs by only tightening the variation around the major contributor to the impedance, the resistor. This is the least expensive part of the circuit and adds only 12 cents to the cost.

We will now approach this sensitivity analysis with an analytical method based on a statistical technique called analysis of variance or ANOVA for short. In this application we will perform an analysis of the sources of variation using the ANOVA calculations. ANOVA is not any more complex than the computation of a SD and can be accomplished with a pocket calculator that has statistical functions. The basic idea of this analysis is to determine where the overall variation is located by computing the components of the total. Let's look at the data for this impedance design in Table 6-3. The SD of the nine impedance values is 4.72, but where did this variation come from? It came from the nine values. Where did these nine values come from? Each one came from a different combination of frequency, capacitance, and resistance. So it must have been the functional influence of these three components that drove the impedance to vary. Now for the best part. The system of variation in these three components is due to an experimental design. What is systematically designed may be systematically decomposed or analyzed! The ANOVA technique is ideal for this purpose.

Table 6-3.

Frequency	Capac.	Resis.	Impedance
216	16	543.5	545.4571
216	20	549.0	550.2344
216	24	554.5	555.3390
240	16	549.0	550.5619
240	20	554.5	555.4801
240	24	543.5	544.2116
264	16	554.5	555.7683
264	20	543.5	544.3450
264	24	549.0	549.5740

Average Impedance = 550.1
Standard Deviation = 4.72

We will begin the analysis by initially ignoring the three potential contributors to the variation in impedance and compute only the total variation (as if there were no individual contributors). This is a merely a recalculation of the variation metric (the SD) that has forced us into this analysis. We will not calculate the SD, but a metric related to the SD. The metric is called the *sum of squares,* and it is additive, whereas the SD is not additive. Since we are seeking the additive components of the variation, the SD is not an appropriate metric.

The calculation of the sum of squares is almost like calculating the SD. The formula is as follows:

$$\text{Sum of squares} = \frac{\text{sum } Y_i^2}{\#\text{ obs. in } Y_i} - \frac{(\text{sum } Y_i)^2}{n}, \qquad (6\text{-}1)$$

where the Y_i's are the responses and n is the number of responses.

If we divide the sum of squares by the number of observations minus one, and then take the square root of this result, we will have the SD. For the impedance example, the sum of the squares of the Y comes from squaring each observation and adding these nine values. The sum of Y_i, quantity squared, comes from adding up all the impedance values and then squaring this sum. The numerical results are shown below.

$$\text{Sum } Y_i^2 = 2723749; \text{ sum } Y_i = 4950.9732; n = 9.$$

Now we will complete the calculations

$$\underset{\text{(Total)}}{\text{Sum of squares}} = 2723749 - \frac{(4950.9732)^2}{9} = 178.3796. \qquad (6\text{-}2)$$

Next we will isolate that part of the total sum of squares that is due to the first factor, frequency. Table 6-4 has arranged the responses to show the three separate levels in groups of three. We will look at how the levels of frequency influenced the impedance by taking the total of the impedance for each group as shown in Table 6-4 and by applying the sums-of-squares formula. We will essentially be looking at how much variation the frequency adds to the total when it is held in isolation.

$$\underset{\text{(frequency)}}{\text{Sum of squares}} = \frac{1651.0305^2 + 1650.2536^2 + 1649.6873^2}{3} - \frac{(4950.9732)^2}{9}. \qquad (6\text{-}3)$$

$$\text{Sum of squares} = .30316.$$

Table 6-4.

Frequency	Capac.	Resis.	Impedance	
216	16	543.5	545.4571	
216	20	549.0	550.2344	Level 1 (216hz)
216	24	554.5	555.3390	sum = 1651.0305
240 #	16	549.0	550.5619	#
240 #	20	554.5	555.4801	# Level 2 (240hz)
240 #	24	543.5	544.2116	# sum = 1650.2536
264 *	16	554.5	555.7683	*
264 *	20	543.5	544.3450	* Level 3 (264hz)
264 *	24	549.0	549.5740	* sum = 1649.6873

In expression 6-3 we divide the sum of the squares of the observations (sum Y_i^2) by three since there were three individual observations involved with each of the totals. In this calculation, we are essentially quantifying the effect of the three levels of the frequency on the average impedance. However, to avoid round-off mistakes in the calculation, we utilize the totals rather than the averages, and then perform the division step last. (A good numerical practice that should be followed in any calculation is to do the division last to avoid round-off errors.)

We will now repeat the calculation of the sum of squares, but this time we will regroup the impedance data according to the pattern of change that corresponds to the capacitance (Capac.) factor. Table 6-5 shows this regrouping and the corresponding sums for each level.

$$\text{Sum of squares (capacitance)} = \frac{1651.7873^2 + 1650.0595^2 + 1649.1246^2}{3} - \frac{(4950.9732)^2}{9}, \qquad (6\text{-}4)$$

Sum of squares = 1.85153.

In expression 6-4 we have once again taken the totals for each level of the factor under study (capacitance in this case); squared these totals and divided by the number of individual observations in the totals we have squared; and then subtracted the grand total squared, divided by the total

Table 6-5.

Frequency	Capac.	Resis.	Impedance	
216	16|	543.5	545.4571|	
216	20#	549.0	550.2344#	| Level 1 (16mfd)
216	24*	554.5	555.3390*	sum = 1651.7873
240	16|	549.0	550.5619|	
240	20#	554.5	555.4801#	# Level 2 (20mfd)
240	24*	543.5	544.2116*	sum = 1650.0595
264	16|	554.5	555.7683|	
264	20#	543.5	544.3450#	* Level 3 (24mfd)
264	24*	549.0	549.5740*	sum = 1649.1246

Table 6-6.

Frequency	Capac.	Resis.	Impedance	
216	16	543.5|	545.4571|	
216	20	549.0#	550.2344#	| Level 1 (543.5 ohms)
216	24	554.5*	555.3390*	sum = 1634.0137
240	16	549.0#	550.5619#	
240	20	554.5*	555.4801*	# Level 2 (549 ohms)
240	24	543.5|	544.2116|	sum = 1650.3703
264	16	554.5*	555.7683*	
264	20	543.5|	544.3450|	* Level 3 (554.5 ohms)
264	24	549.0#	549.5740#	sum = 1666.5874

number of observations from this result. Both the frequency sum of squares as well as the capacitance sum of squares are quite small compared with the total sum of squares. The next source of variation will prove to be different.

$$\text{Sum of squares (resistance)} = \frac{1634.0137^2 + 1650.3703^2 + 1666.5874^2}{3} - \frac{(4950.9732)^2}{9}, \qquad (6\text{-}5)$$

Sum of squares = 176.84208.

The sum of squares for the resistance is substantially greater than that for the frequency and the capacitance. Since the resistance sum of squares is nearly as large as the total sum of squares (which is directly related to the total variation of the entire circuit), the resistor must be the major contributor to the variation! We may summarize our calculations in an ANOVA table. To build such a table, we set up columns for the name for the *source* of variation, the *sum of squares,* the degrees of freedom (df), and the *% contribution.* Table 6-7 shows the summary ANOVA for this impedance data.

Now for the action to reduce the variation in our product. Since the resistor is the only real contributor to the overall variation in the impedance, and the customer wants the variation reduced to half of what it is now, we simply need to reduce the variation in the resistor to one-half of its current tolerance to accomplish our goal. Table 6-8 shows the new design that will confirm this analysis. When we compute the variation of this new engineering design, we obtain an acceptable SD of 2.33, which is exactly what the customer required. Now, this is quality engineering by design! We have accomplished our goal and in doing so have come to the final design without reworking engineering effort.

Now that we have achieved the goal of the "right" variation, we may ask why was the resistor the major contributor to this variation? If we look at the resistance values in Table 6-8, we can see the reason. The resistor accounts for over 99% of the total impedance! The contribution of the two other components is minor and therefore they cannot contribute to the

Table 6-7.

ANOVA (ANalysis Of VAriation)

SOURCE	SUM OF SQUARES	df	% CONTRIBUTION
FREQUENCY	.30316	2	.17%
CAPACITANCE	1.85153	2	.68%
RESISTANCE	176.84208	2	99.14%
RESIDUAL		2	.01%
TOTAL	178.3796	8	

Table 6-8.

Frequency	Capac.	Resis.	Impedance
216	16	546.5	548.2376
216	20	549.0	550.2344
216	24	551.5	552.5536
240	16	549.0	550.5619
240	20	551.5	552.6955
240	24	546.5	546.9983
264	16	551.5	552.9851
264	20	546.5	547.1310
264	24	549.0	549.5740

standard deviation of the impedance = 2.33

variation. We pointed this out in Chapter 5 as the physical reason that the parameter design performed its function. This is the same physical reason that the tolerance design has brought us to this conclusion. The statistical analysis merely guided us to this same conclusion.

While the above example is easily seen through, the next example will illustrate the power of the tolerance design and analysis that involves a more complex problem where we do not have a mathematical expression. However, before we move to a more comprehensive application of the tolerance design and analysis, we need to clear up a minor consideration about the distribution of the output of the tolerance design and the interpretation of the variation in this distribution.

In Figure 6-3 we have illustrated how the tolerance design has produced the results that have led us to the desired outcome. We have utilized a designed experiment to expedite the generation of the impedance data and, more importantly, to allow a systematic analysis of this data to determine the quality-sensitive components. However, the designed experiment has a structure, and each of the levels or conditions for each factor appears systematically. We need this system for the analysis, but such a systematic variation in the factors will never take place in reality. We would expect a random, possibly normal distribution of conditions in the building of the impedance circuit. In general, we would not expect the systematic pattern of input factors as shown in Figure 6-3, but rather a normal distribution of the inputs. These normally distributed inputs would

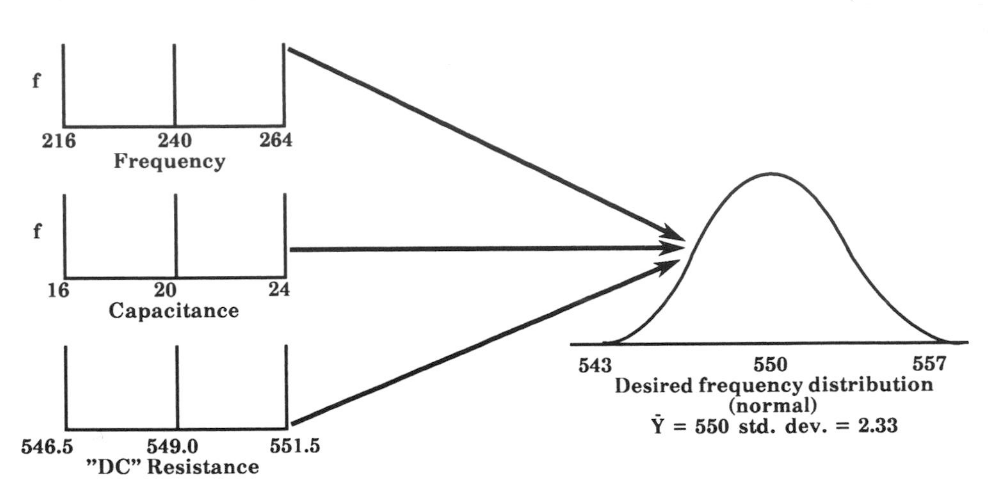

Figure 6-3.

then lead to the desired normal distribution of the output. The input distributions in the impedance example are not normal, but are actually rectangular or uniform distributions.

The crux of the problem lies in the conflict between expediting the experiment with a small number of runs and obtaining the correct output distribution. Fortunately, there need be no compromise to resolve this conflict, for a simple transform of the input distribution levels will produce statistics that appear in the output distribution as if they had come from a normal distribution. Figure 6-4 illustrates how this is accomplished.

In each of the input distributions, the levels are set just a bit wider than the SD for that factor. For a three-level design, the factor 1.2247 is multiplied by the SD to obtain the actual levels in the tolerance design. Appendix 1 shows the logic of the mathematics behind this transform. The following is an example of the application of this transform.

Let's say that the frequency generator from the impedance example may be purchased from an electronics supplier. In the catalogue, the manufacturer quotes a tolerance on the frequency output as ±25%. Without further clarification, we could interpret this tolerance to mean ±3 SD around the average frequency. (It would be best to ask the manufacturer what is actually meant by "tolerance" in each situation. It would be

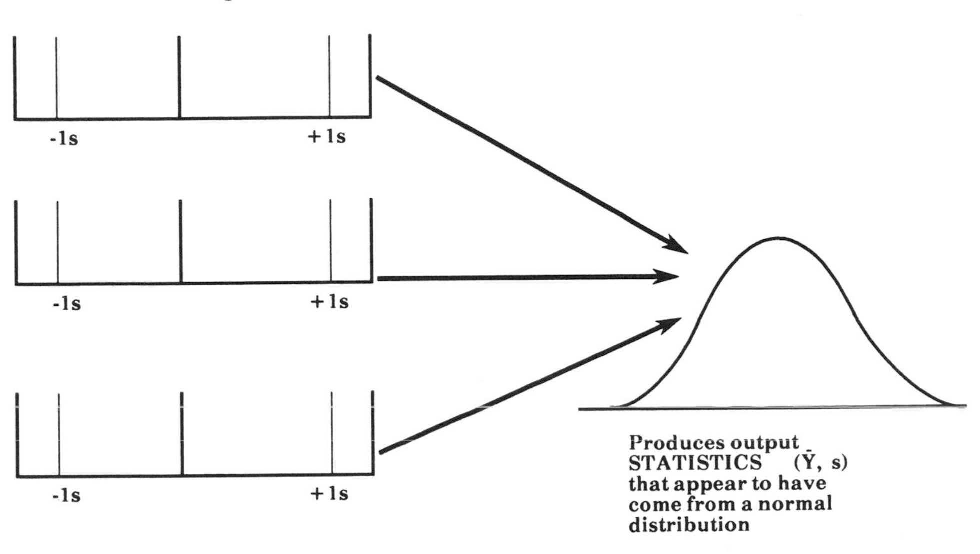

Figure 6-4.

much better if the tolerance were stated in units of SD. Then we would not need to guess what was meant!) If we adhere to our assumption that 25% is 3 SD, then 1 SD is 8.33 (25/3). We now multiply the 8.33 by 1.2247 to obtain 10.2% (which we rounded to 10% in the impedance example).

In general, we obtain a value for the SD of the input distribution and then multiply by 1.2247 to set the levels of that factor in the tolerance design. If we only use the SD without this "expansion factor," we will underestimate the SD of the output distribution. If we use 3 SD as the set points in the experimental design, we will vastly overestimate the output distribution's SD.

To obtain the SD of the input distribution, we must either rely on published information from a supplier, determine its value by measurement, or run a process capability study on the characteristic if it is under our design authority. Table 6-9 is a summary of these sources.

Now that we have established a method to produce statistics from our tolerance design that may be interpreted to have emerged from a normal distribution, we will look at a more complex example of this powerful sensitivity analysis in action.

Table 6-9.

Sources of Input Variation Information.

- Published vendor information (specification sheets)
- Direct measurement on samples (incoming inspection)
- Process capability studies

Table 6-10 is a tolerance design for an ink used in an ink-jet printer. The parameter design has determined the optimal settings for each of the five factors under our design authority. Table 6-11 lists the factors, the

Table 6-10.

Suften	Visc	P size	P count	Drive	Storage	Response #satellites
28.8	3.8	1.2	9.6	.75	3.3	37.0
28.8	3.8	2.0	10.0	1.0	4.5	39.9
28.8	3.8	2.8	10.4	1.25	5.6	41.3
28.8	5.0	1.2	10.0	1.0	5.6	40.7
28.8	5.0	2.0	10.4	1.25	3.3	39.1
28.8	5.0	2.8	9.6	.75	4.5	43.4
28.8	6.2	1.2	10.4	1.25	4.5	39.7
28.8	6.2	2.0	9.6	.75	5.6	45.6
28.8	6.2	2.8	10.0	1.0	3.3	42.5
30.0	3.8	1.2	10.0	1.25	4.5	35.8
30.0	3.8	2.0	10.4	.75	5.6	45.3
30.0	3.8	2.8	9.6	1.0	3.3	38.5
30.0	5.0	1.2	10.4	.75	3.3	40.1
30.0	5.0	2.0	9.6	1.0	4.5	39.3
30.0	5.0	2.8	10.0	1.25	5.6	40.7
30.0	6.2	1.2	9.6	1.0	5.6	40.7
30.0	6.2	2.0	10.0	1.25	3.3	39.1
30.0	6.2	2.8	10.4	.75	4.5	47.1
31.2	3.8	1.2	10.4	1.0	5.6	41.0
31.2	3.8	2.0	9.6	1.25	3.3	35.7
31.2	3.8	2.8	10.0	.75	4.5	43.7
31.2	5.0	1.2	9.6	1.25	4.5	35.8
31.2	5.0	2.0	10.0	.75	5.6	45.3
31.2	5.0	2.8	10.4	1.0	3.3	42.2
31.2	6.2	1.2	10.0	.75	3.3	40.7
31.2	6.2	2.0	10.4	1.0	4.5	43.6
31.2	6.2	2.8	9.6	1.25	5.6	41.3

Table 6-11.

Factor	Optimal Midpoint	Low Cost Tolerance	Std. Dev.	Std. Dev. x 1.2247*
Surface Tension	30 dynes/cm.^{2}	3 d/cm.^{2}	1 d/cm.^{2}	1.2 d/cm.^{2}
Viscosity	5 cps	3 cps	1 cps	1.2 cps
Particle Size	2 micrometers(u)	2 u	.667 u	.8 u
Particle Count	10	1	.333	.4
Drive	1.0 amps	0.6 amps	.2 amps	.25 amps
Storage	4.5 weeks	2.9 weeks	.95 week	1.15 week

*rounded to a reasonable value

midpoint setting, the low-cost three sigma tolerances, as well as the SDs and the variation around the midpoint to establish the levels in the three-level design used in this tolerance analysis.

An L27 design (Table 6-10) was used as the design configuration for this tolerance analysis. The ANOVA of the response from this design shows that there is more than just one quality- sensitive component. In the ANOVA (Table 6-12) we find that viscosity is an 11% contributor, particle size is a 21% contributor, particle count is an 11% contributor, and drive is a 37% contributor. While storage contributes 17% to the variation, it is an outside noise factor that we have no control over and therefore may not reduce its tolerance. It is very much like the residual or uncontrolled variation (1.4% in this case) and cannot be held tightly to reduce the overall variation in the response.

In the previous tolerance analysis with the impedance, there was only one component (the resistor) that required tolerance tightening, since it was only the major (99%) contributor to the overall variation of the im-

Table 6-12.

ANOVA SOURCE	SUM OF SQUARES	df	% CONTRIBUTION	
SURFACE TENSION	0.5	2	0.2	
VISCOSITY	27.5	2	11.5	QUALITY
PARTICLE SIZE	51.1	2	21.5	SENSITIVE
PARTICLE COUNT	26.9	2	11.4	
DRIVE	87.4	2	36.8	COMPONENTS
STORAGE	40.5	2	17.1	
RESIDUAL	3.2	14	1.4	
TOTAL	237.3	26		

pedance. However, in this example, we have four components that contribute to the overall variation. When there is only one component, it is obvious that it is the only one to be tightened. How do we react when there is more than one critical component? In this situation there is a greater degree of complication, but there is also a commensurate degree of flexibility added to the design of the product. This flexibility comes from our ability to write an equation that will allow us to determine the level of tolerance tightening on the quality-sensitive components. We will now examine the rationale behind this equation.

First, we must determine the allowable variation in our product characteristic. In the limit, this variation is zero! However, we are realistic and understand that while our goal is to reduce the variation to zero, we will tolerate the level of variation that will produce an acceptably low loss for our situation. It is the loss function that drives the tolerance. Figure 6-5 shows the loss function for our particular problem. When we encounter ten satellites around a drop of ink, the resulting image becomes fuzzy enough to trigger a service call that costs $120. This gives us a k value of $1.20.

The current SD is 3 and would look like the distribution (*) superimposed on the loss function in Figure 6-6. This distribution causes an expected loss of $5.40. (Note: For a one-sided loss function, the loss only takes places for the intersecting areas of the loss curve and the characteristics distribution. In this case we have one-half the EL.)

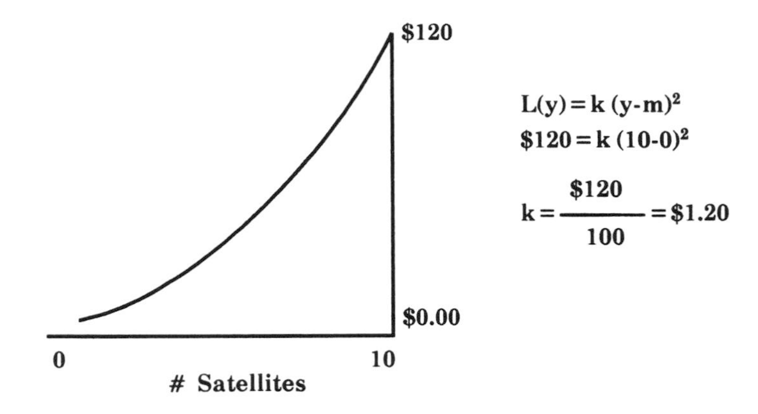

Figure 6-5.

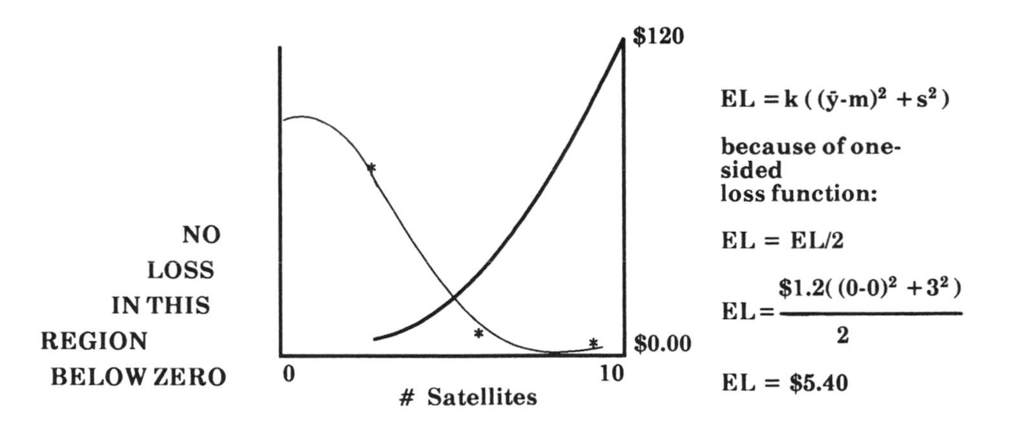

Figure 6-6.

We would like to reduce our loss by \$3 and have a resulting loss of only \$2.40. To do so, we must reduce the spread (or the SD [s]) of the satellite distribution. To determine the new required SD, we solve the expected loss formula backwards and solve for s, given the loss.

$$EL = \frac{k[(\bar{y} - m)^2 + s^2]}{2}. \tag{6-6}$$

$$\$2.40 = \frac{\$1.20[(0 - 0)^2 + s^2]}{2}. \tag{6-7}$$

$$\$4.80 = \$1.20(s^2). \tag{6-8}$$

$$\frac{\$4.80}{\$1.20} = s^2.$$

$$4 = s^2.$$

$$s = 2.$$

The new, tightened distribution is superimposed on the loss function in Figure 6-7. This narrower distribution accomplishes our goal of lower loss, but we must now determine *how* to establish the lower variation satellite distribution. Of course, we realize that the satellite formation is a

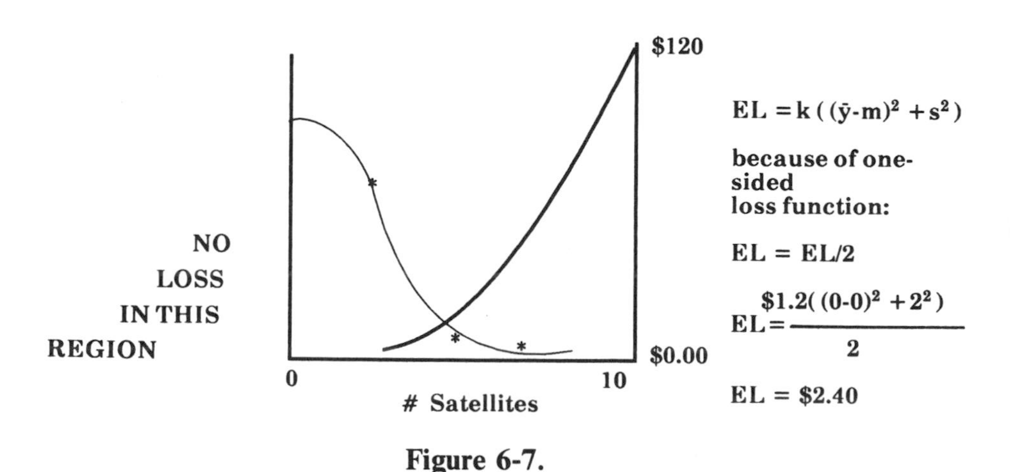

Figure 6-7.

function of the formulation of our ink, the drive used to accelerate the particles toward the paper, and, unfortunately, the storage time of the ink. We will now be able to link these factors to the satellite distribution and determine new, tighter tolerances for the factors under our control. We are essentially observing the transmission of variation through our ink-jet system.

The analysis of the variation in our problem that was found in Table 6-12 is summarized in Table 6-13. We will reduce the tolerance of only the quality-sensitive components that are under our design authority. Each component will be weighted according to its percent contribution to the overall variation. We will reduce the variation in the components to achieve an overall SD of 2 (as computed above) from the current SD of 3.

Since SDs are not additive (or muliplicative), we will work with variances (the square of the SD). We must reduce our current variance of 9 to a variance of 4. We need a variance that is four-ninths (.444) of the current value to achieve our lower loss goal. The only way we can accomplish this reduction is to reduce the contributors by tightening the variation around them. This concept may be expressed in an equation. In this equation we will state our goal (in this case, .444) and propose reductions in the quality-sensitive components that will achieve this goal. Expression 6-9 is the equation for our particular situation.

Table 6-13.

SOURCE	% CONTRIBUTION	
VISCOSITY	11.5	QUALITY
PARTICLE SIZE	21.5	SENSITIVE
PARTICLE COUNT	11.4	
DRIVE	36.8	COMPONENTS
STORAGE	17.1	OUTSIDE NOISE
RESIDUAL (includes surface tension)	1.7	

TOTAL sum of squares = 237.3; 26 df; s = 3.02

$$.444 = [(VIS)^2 \cdot .115 + (PSZ)^2 \cdot .215 + (PTC)^2 \cdot .114 + (DRV)^2 \cdot .368 + .188], \quad (6\text{-}9)$$

where VIS = fractional reduction of viscosity, PSZ = fractional reduction of particle size, PCT = fractional reduction of particle count, and DRV is the fractional reduction of drive.

The beauty as well as the difficulty with this equation is the fact that there are many different solutions for it. This is an advantage, since we are able to try different approaches to reducing the variance and to utilize the most economical approach. The difficulty comes from the fact that we have only one equation with four unknowns! A key to overcoming this problem is to remember the basic concept behind an equation: We simply must find values of the unknowns (on the right side) that will make the right side equal to the left side.

The first operation we must perform with our equation is to remove the .188 residual from both sides of the equation. Since the residual includes factors or effects unknown to us as well as the noise factor that we cannot control, we are unable to make a reduction. Of course, if the residual becomes bigger than the required variance reduction, we have an impossible problem, and we either must settle for a more variable product (with higher loss) or search for a better system design. In most cases, if the product is unique, we will put that product on the market despite its higher variation, while working on a solution to the problem. The Japanese follow this practice quite often and cut in changes as they become available to make product improvements. This is an "earn while you learn" philosophy.

After removing the residual contribution, our equation looks like this:

$$.256 = [(VIS)^2 \cdot .115 + (PSZ)^2 \cdot .215 + (PTC)^2 \cdot .114 + (DRV)^2 \cdot .368]. \quad (6\text{-}10)$$

Current tolerance:	3 cps	2 μ	1 particle	.6 A

Now we need to examine the four components in our equation and to determine the relative ease, cost, and feasibility of tightening the tolerances on each. The current tolerances are reproduced just below expression 6-10. We realize that it is necessary to cut these tolerances in order to achieve the overall tighter level of variation in the final product characteristic (number of satellites), since these components transmit variation to the response.

Realizing that there are many solutions to expression 6-10, let us try the following set of reductions. We will enter each component's reduction sequentially. Suppose that it is possible to reduce the variation of the viscosity (VIS) to a level that is 50% (or .50) of the current tolerance. This would impose a new tolerance on the viscosity of 1.5 cps, or a SD of .5 cps. Entering the .50 fractional reduction into the equation and subtracting the resulting value from both sides, we obtain expression 6.11c.

$$.256 = [(.50)^2 \cdot .115 + (PSZ)^2 \cdot .215 + (PTC)^2 \cdot .114 + (DRV)^2 \cdot .368]. \qquad (6\text{-}11a)$$

$$.256 = [.029 + (PSZ)^2 \cdot .215 + (PTC)^2 \cdot .114 + (DRV)^2 \cdot .368]. \qquad (6\text{-}11b)$$

$$.227 = [(PSZ)^2 \cdot .215 + (PTC)^2 \cdot .114 + (DRV)^2 \cdot .368]. \qquad (6\text{-}11c)$$

We will propose a fractional reduction to .60 of the current level of variation in particle size (PSZ), as well as a reduction to .40 of the current level of variation in particle count (PTC). The last component, drive (DRV) must now make up the remainder, as shown in expression 6.12f.

$$.227 = [(.60)^2 \cdot .215 + (.40)^2 \cdot .114 + (DRV)^2 \cdot .368]. \qquad (6\text{-}12a)$$

$$.227 = [0.77 + .018 + (DRV)^2 \cdot .368]. \qquad (6\text{-}12b)$$

$$.132 = (DRV)^2 \cdot .368. \qquad (6\text{-}12c)$$

$$\frac{.132}{.368} = (DRV)^2. \qquad (6\text{-}12d)$$

$$.3587 = (DRV)^2. \qquad (6\text{-}12e)$$

$$.60 = DRV. \qquad (6\text{-}12f)$$

Table 6-14 is a summary of the reductions in variation for each component and the new tolerance associated with each component.

Table 6-14.

COMPONENT	ORIGINAL CONTRIBUTION	FRACTIONAL REDUCTION	CURRENT TOLERANCE	NEW TOLERANCE
VIS	11.50%	0.50	3.00000	1.50000
PSZ	21.50%	0.60	2.00000	1.20000
PTC	11.40%	0.40	1.00000	0.40000
DRV	36.80%	0.60	0.60000	0.35942

Table 6-15. Alternative Tolerance Reduction.

The Current Std. Dev. Is: 3 The Required Std. Dev.Is: 2

You Have A Residual Of: 18.80%

The Response's Variation Must Be Reduced To 0.256 Of Its Current Value

COMPONENT	ORIGINAL CONTRIBUTION	FRACTIONAL REDUCTION	CURRENT TOLERANCE	NEW TOLERANCE
VIS	11.50%	0.40	3.00000	1.20000
PSZ	21.50%	0.50	2.00000	1.00000
PTC	11.40%	1.15	1.00000	1.15156
DRV	36.80%	0.30	0.60000	0.18000

It is possible to try many other reduction combinations. Table 6-15 shows how trade-offs may be made between components to actually *increase* a tolerance. In this case, we have tightened the viscosity and the particle size tolerances slightly and tightened the drive tolerance to one-half of the level in Table 6-14, which in turn is one-third of the original. This allows the particle count tolerance to be opened up, thus saving the cost of holding this factor at a tight tolerance. Of course, we have increased the cost of holding the other factors at tighter levels, but it is possible to compare the costs of various tolerance reduction scenarios and to use the least costly one. This is truly quality engineering, since we attain the proper level of quality at the least cost.

PROBLEMS FOR CHAPTER 6

1. Propose a tolerance design for the following ink formulation used in an ink-jet computer printer. This is the same situation as described in problem #1 of Chapter 5, and the goal of the effort is to reduce the blur in the image made by the ink-jet printer by reducing the number of free particles of pigment (carbon black) that surround the ink droplets. The parameter design has been run and the levels of the factors that put the response on target with low variation are listed below along with their low-cost variation.

Factors	Level	Low-cost variation
Surface tension	30 gyn/cm^2	SD = 3 dyn/cm^2
Viscosity	5 centipoise	SD = 1 centipoise
Pigment particle size	2 μm	SD = .5 μm
Number of particles	10 per μl	SD = 1 particle
Printer drive power	1.0 A	SD = .05 A
Storage	1, 13, 26 weeks	Outside noise

Storage of ink before use is an outside noise factor that could range between 1 and 26 weeks.

2. The charge voltage in an electrostatic copier depends on the position of the charging wire. The quality of the copy depends on the charge voltage. If the charge is too low, the image will be light. If the charge is too high, the image will be very black, but there will be an accompanying high level of background, which is a gray mist all over the copy. The mechanical engineer must position the bracket that holds the wire to impart the correct charge consistently. Through a parameter design he has accomplished part of this goal, but the variation is still to high. The set-point values for each of the factors is listed below along with the SD of these settings. The contribution to the overall variation is also listed for each factor. Determine the rational reduction of the variation based on this data and recommend the steps required to bring this charging system to a level of high quality.

Factors	Set point	Low-cost SD	Percent contribution
Bend angle of the bracket	89.5°	.5°	24.4%
Thickness of metal	2.0 mm	.05 mm	0.3%
Hole location (from reference points)			
Unthreaded	15.05 mm	.05 mm	0.2%
Threaded	15.0 mm	.05 mm	0.1%
Torque on fasteners	110 in. oz.	2 in. oz.	0.0%

Outside Noise Factors

Relative humidity	30–70%		0.8%
Current output of power supply	20 μA	.5 μA	74.2%

The current SD of the output voltage is 23.
The required SD of the output voltage is 5.

3. Propose a tolerance design for the following automotive optimization. The engineering team has attained a high value of miles per gallon (mpg), but the variation from car to car is still too great due to the following possible manufacturing tolerances. Your tolerance design will address the identification of the quality-sensitive component (s).

Factors	Set point	Low-cost SD
Spark plug gap	35 thousandths	.5 thousandths
Carb. adjustment	4 turns	.25 turn
Fuel octane	87	1
Spark advance	20°BTDC	1°BTDC

4. Determine a rational reduction of variation from the following chemical engineering experiment. Molecular weight is the response. The target value of 200,000 has been achieved, but the variation is too great.

Factors	Set point	Low-cost SD	Percent contribution
Temperature	200°C	10°C	30.2%
Pressure	20 psi	1 psi	38.5%
Reaction time	20 min	.5 min	10.2%
Catalyst concentration	.5%	.05%	8.8%
Residual (unknown)			12.2%

The current SD is 40,000. The required SD is 20,000.

APPENDIX 6-1

The rationale behind the 1.2247 correction factor for the production of normal distribution statistics from input distributions that are uniform in a tolerance design situation is shown below.

In Figure A6-1, we have a uniform distribution with three points, $\mu - a$, μ, and $\mu + a$. The definition of the variance is:

$$\sigma^2 = \frac{\sum (x - \mu)^2}{N}. \tag{A6-1}$$

We have three x's in our population, which we put into the variance formula:

$$\sigma^2 = \frac{((\mu - a) - \mu)^2 + (\mu - \mu)^2 + ((\mu + a) - \mu)^2}{3}. \tag{A6-2}$$

$$\sigma^2 = \frac{(-a)^2 + (a)^2}{3}. \tag{A6-3}$$

$$\sigma^2 = \frac{2a^2}{3}. \tag{A6-4}$$

$$a = \sqrt{\frac{3}{2}}\,\sigma. \tag{A6-5}$$

$$a = 1.2247\,\sigma. \tag{A6-6}$$

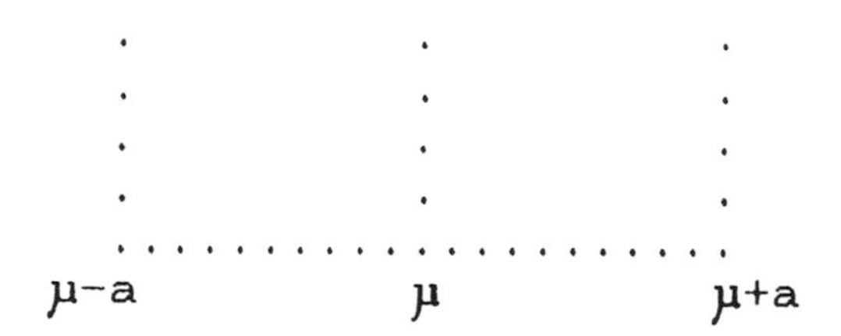

Figure A6-1

In Figure A6-2, we have superimposed a uniform distribution with a width of ±1 on a normal distribution with a width of ±3 SD. Notice how the area of the uniform distribution extends above the area of the normal (symbolized with *). Also notice how the normal has area beyond the uniform (symbolized with #). If we measure the excess area found in the normal and compare this excess area (#) with the excess area (*) that the uniform has above the normal, there is more excess in the normal than there is in the uniform if the uniform is set at ±1 SD.

In Figure A6-3, we have set the width of the uniform at 1.2247 SD from the midpoint. Now when we compare the excess area of the uniform with the excess area of the normal, there is a match. By having area in the input uniform distribution equivalent to a ±3 SD normal, we are able to produce statistics from the tolerance design that utilizes three-level experimental structure statistics that appear to have come from input distributions with a normal shape. Since so many of our input distribution characteristics are normally distributed, this transform is an excellent way to accomplish the tolerance design sensitivity analysis with a minimum of experimental runs.

It is possible to compute the factors for other experimental design configurations using the method shown in expressions A6-1 through A6-6. It is left for the student to confirm that the factor for a two-level design is 1.0. Note that for this analysis method to work, it is necessary that all factors are set at the same number of levels.

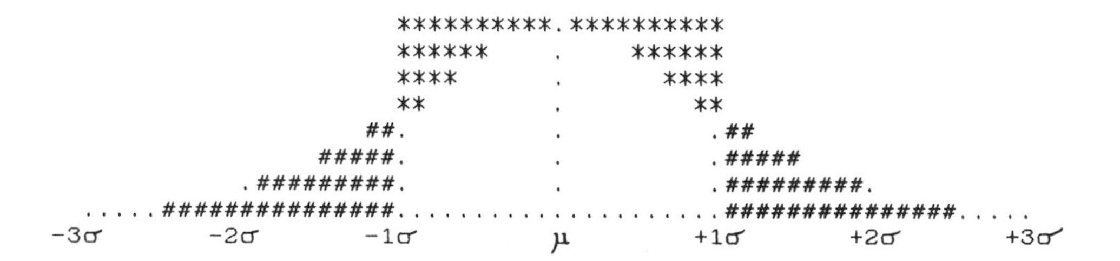

Figure A6-2

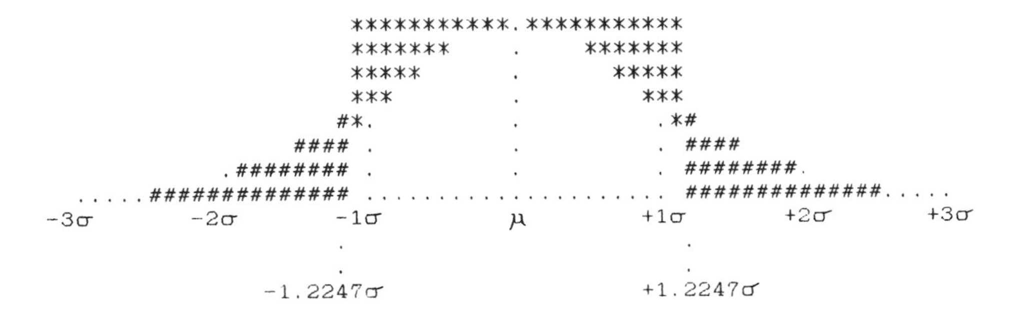

Figure A6-3

Consequences of Not Using the $\sqrt{3/2}$ Transform

What if we do not use the above transform to make three-level input designs produce statistics that look like they came from "normal" input distributions? Figure A6-4 shows a 3σ normal distribution (#) and a superimposed uniform distribution (*) with levels set at the -3σ point, midpoint, and at the $+3\sigma$ point.

The problem with setting these levels for the tolerance design is obvious. The probability at the + or -3σ points in a normal distribution is practically nill. However, in the setting of the levels of our tolerance design (which is always a uniform distribution) at $\pm 3\sigma$, we have given the extremes the same weight (or probability) that the midpoint position has! This would in effect produce an output variation that is vastly inflated. In an early application of tolerance design to an industrial problem, the engineer who had been accustomed to providing $\pm 3\sigma$ tolerance to his Monte-Carlo simulation program did the same when he set up a tolerance design. He was forced to the tolerance design, because the mathematical model he was using to predict the outcome required one-half hour of computer time (on a mainframe!) for just one solution of the model. A Monte Carlo would have required at least 1000 such solutions, and this would have consumed more computer budget than was available. However, in setting the levels at $\pm 3\sigma$, the engineer overestimated the output variation and believed that the product could not be manufactured! He recommended that the product be cancelled as a result of the faulty setting of levels in the tolerance design!

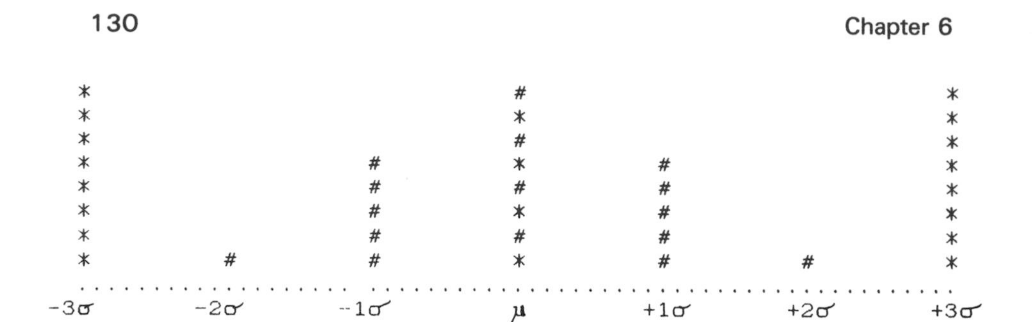

Figure A6-4

On the other hand, if we set the levels at only ±1σ from the midpoint, we will underestimate the variation. In this case, when the product goes into production, we will have a greater loss due to the higher variation.

Therefore, the recommendation is to take the ±3σ tolerance, divide it by 3 (to obtain the 1σ value) and then multiply this by 1.2247 to build the three-level tolerance design. Remember, this transform is only useful in three-level designs. Other transforms must be determined for other design configurations.

Appendix Reference

D'Errico, J.R., Zaino, N.A.: Statistical tolerancing using a modification of Taguchi's methods. *Proceedings QED 87.* Rochester Institute of Technology, Rochester, NY, 1988.

APPENDIX 6-2

ACCOUNTING FOR VARIATION

In this chapter, we used a statistical method known as the ANOVA. We used ANOVA to identify and quantify the sources of variation as a function of the factors in our experiment. This appendix will show how this ANOVA concepts works based only on the simple understanding of the calculation of SD.

Before we begin, there are two basic concepts.

1. SDs are not additive, but the quantity that leads to the calculation of the SD is always additive.
2. The quantity that leads to the calculation of SD(s) is called the *sum of squares* (around the mean). We will use the symbol SS for this quantity.

$$\text{recall:} \quad s = \sqrt{\frac{\text{sum } X_i^2 - \dfrac{(\text{sum } X_i)^2}{n}}{n - 1}}.$$

The numerator within the radical is the sum of squares.

$$SS = \text{sum } X_i^2 - \frac{(\text{sum } X_i)^2}{n}.$$

The following 32 observations are sent to us for analysis. Our first instinct is to plot the data in a frequency histogram and to compute the average and SD.

Table A6(2)-1.

2	3	4	6	3	4	5	7	
3	4	6	8	4	5	7	9	$\bar{X}$ = 6.5
4	5	8	10	5	6	9	11	s = 2.87
5	6	10	12	6	7	11	13	

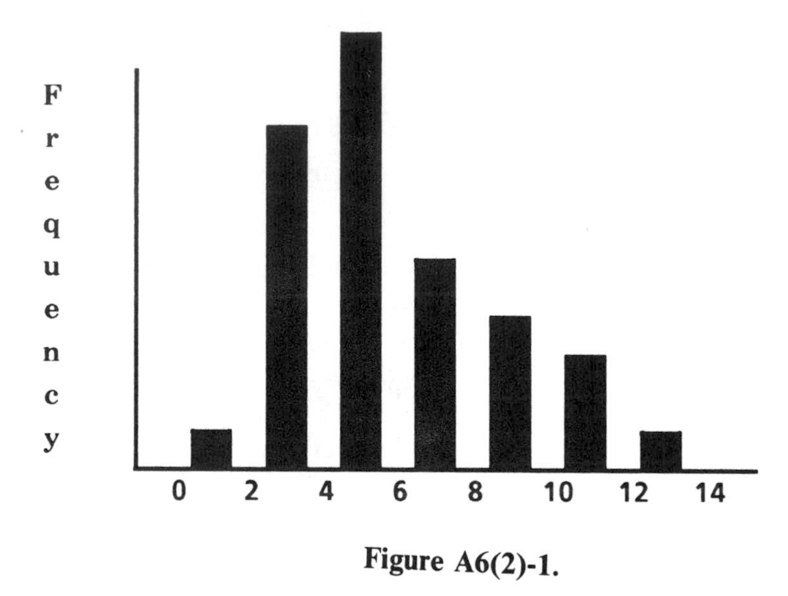

Figure A6(2)-1.

The distribution appears to be skewed (not symmetrical), and we may be interested in knowing what caused the variation to take place. We dig a bit farther into the experiment and find that what we have came from an eight-run factorial design. When we rearrange the data in Table A6(2)-2 we begin to see some of the pattern of change. From this insight into the pattern of change, we are able to account for some of the sources of the overall variation.

Now we will express the variation in its additive form (SS). The overall or total SS is 256. (Divide this by 31 and then take the square root and you have the SD!)

We ask ourselves how much influence does each of the factors (A, B, C) have on the response. If there is an influence by a factor, we will see such a pattern if we compare the clustering of the data at either level of the factor. A very simple cluster would be the total. This clustering is shown in Table A6(2)−3. We may now use the cluster points (totals) to find the variation (expressed as SS) for each factor, as shown in Table A6(2)−3.

Since sums of squares are additive, we may find how much of the total variation was due to all of the factors by adding the SS_A with the SS_B and the SS_C. This sums to 154, but the total SS was 256. So there is still an unac-

Table A6(2)-2.

A	B	C	Responses			
1	1	1	2	3	4	5
2	1	1	3	4	5	6
1	2	1	4	6	8	10
2	2	1	6	8	10	12
1	1	2	3	4	5	6
2	1	2	4	5	6	7
1	2	2	5	7	9	11
2	2	2	7	9	11	13

Table A6(2)-3.

	A(1)	A(2)	B(1)	B(2)	C(1)	C(2)
TOTALS:	92	116	72	136	96	112

$$\frac{92^2 + 116^2}{8} - \frac{208^2}{32} \qquad \frac{72^2 + 136^2}{8} - \frac{208^2}{32} \qquad \frac{96^2 + 112^2}{8} - \frac{208^2}{32}$$

$$SS_A = 18 \qquad SS_B = 128 \qquad SS_C = 8$$

counted SS of 102. This if often called a *residual* or *error* in a typical statistical analysis. If we can find a source or sources for this residual, we will have a better understanding of our process. With the following hint, that search is left for the reader as an exercise. Hint: Each of the four observations came from a designed experiment with two factors (A′ and B′), which had two levels each.

7
Parameter Design—Making the Experimental Effort Viable

In Chapter 5, we explained that an important concept of the engineering quality by design methodology is the control of variation in the product. To understand the source of the variation, we must either have equations or mathematical models of the process that may be exercised with the parameter design or, more realistically, we must run physical experiments to obtain this information empirically. The information we seek for product or process optimization involves both the location characteristic of our responses as well as the variation around these response variables. The basic premise of parameter design is the hypothesis that there is a functional relationship between the controlling factors that we may manipulate and the responses we measure. We also hypothesize that we may compensate for the variation in factors beyond our control (noise factors) by the proper choice of the levels of the factors under our control.

In Figure 7-1, we observe that the slope (or change) in the response is greatest when the controlling factor is set at its high level. Since we are usually unable to specify the noise factor's range, it would be much better in this situation to operate the process at the low level of the controlling factor, since the variation in the response is much less at this point, even though the noise factor is allowed to vary over its entire range. The parameter design produces the understanding that allows us to "immunize" our design to variation in outer noise factors. Taguchi calls this the process of *robustification.*

In Chapter 5 we used the outer array (or noise matrix) to generate the variation that was subsequently measured with the S/N figure of merit. The concept behind this method is actually quite simple. We want to find the level of the controlling factor that will withstand variation best. This robust condition is identified by causing variation to take place and by comparing the magnitude of the variation at different settings of the controlling factors. The outer arrays (or noise matrices) generate variation through systematic changes in the tolerances of the controlling factors (inner noise), as well as the variation induced from sources beyond our engineering authority (outer noise). The parameter design is successful because we take advantage of the nonadditives in the system. The nonadditives take the form of either interactions or second-order curvilinear relationshps. Figure 7-1 illustrates the interaction nonlinear mechanism that is behind parameter design.

Another form of nonadditivity is found in Figure 7-2, where the response to factor X is a quadratic function. A quadratic function also exhibits different slopes, similar to an interaction. Unlike the interaction where there are two factors producing the changing slope, the "slope" in

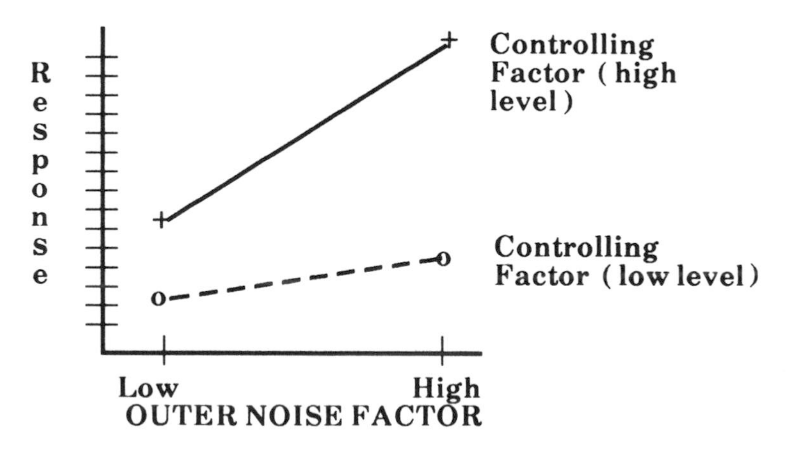

Figure 7-1.

the quadratic changes as a function of the factor itself! In a sense, the quadratic factor is "interacting" with itself to produce different change relationships at different levels of this factor. Whenever there are differential slopes, we will make our selection of functional levels where the slope is minimal, for this setting or level will transmit the least variation to our response. This concept is a very important consideration in the *internal stress method* (ISM), which we shall cover in this chapter.

AN EXAMPLE

First we shall look at the process of parameter design that uses the full inner array (design matrix) and outer array (noise matrix) approach to its solution. To obtain the data for this example, we will use an equation that produces the delivery time of a robotics device. In this way, we will be able to try other methods of parameter design and be able to judge their effectiveness compared with the benchmark inner/outer array method.

While the inner/outer array method illustrates the concepts of the engineering quality by design parameter, design process, it is not the most practical method to accomplish our quality engineering goals, because of the vast number of runs required and the exactness of the experimentation required to make these runs. The basic premise of the parameter design is to have the outer arrays (or noise matrices) cause variation and to find the conditions in the inner array (or design matrix) that will minimize the effect of this noise on our response's variation. In essence,

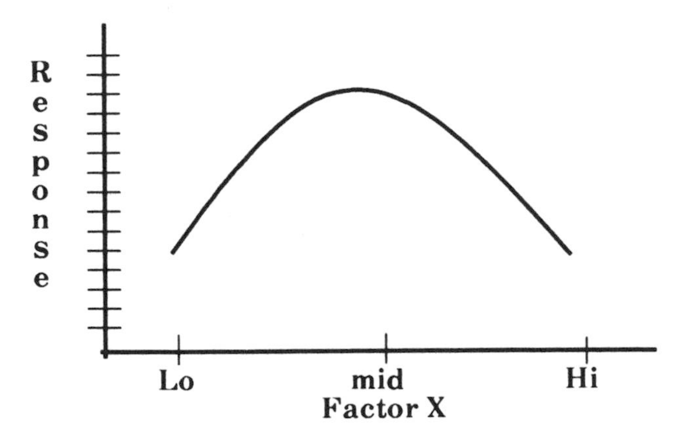

Figure 7-2.

we are performing a sensitivity analysis on the transmission of variation in a process as well as immunizing this process to conditions beyond our control.

We will now design an experimental configuration to study the variation. There are four controlling factors in the robotics equation (found in Appendix 7A) and two noise factors. Therefore, the minimum size design for three levels is an L9. We could also use the L27, but if the number of interactions and their intensity can be considered to be small compared with the single effects, then a substantial saving may be made by using this smaller L9 design. We must realize that all the information generated by this experiment is being used for the single effects. It is, then, a *saturated* design capable of only detecting the single effects.

We must now generate the outer arrays to perturb the midpoint settings with tolerances as well as noise factors. For this part of the parameter design, we must use a design matrix that will accommodate six factors (four controlling factors and two noise factors). We will use the L18, since we do not intend to perform a trend analysis on this part of the experiment and any interactions confounded in this design configuration will not cause any problems. Therefore, each run of the design matrix will trigger 18 observations in the noise matrix. This activity of producing a noise matrix is similar to replicating each run in the design matrix 18 times. However, instead of waiting for the random replicates to generate variation, the outer array makes variation happen by design. We will show the noise matrix (Table 7-2) only for run #1 of the design matrix.

We will now use the experimental design configuration shown in

Table 7-1.

RUN #	Inertia	Motor Ld.	Spring Ld.	Torque	RESPONSE
1	1.7	.20	45	.001	
2	1.7	.25	90	.015	
3	1.7	.30	135	.03	
4	11.5	.20	90	.03	
5	11.5	.25	135	.001	
6	11.5	.30	45	.015	
7	21.4	.20	135	.015	
8	21.4	.25	45	.03	
9	21.4	.30	90	.001	

Table 7-2.

RUN#	Inert	Mtr Ld	Sprg Ld	Torque	Prt Wt	Prt Sz	RESPONSE
1	1.6	.175	41.3	0	8	7	164
2	1.6	.200	45.0	.001	59	8.5	178
3	1.6	.225	48.7	.002	110	11	178
4	1.7	.175	41.3	.001	59	11	219
5	1.7	.200	45.0	.002	110	7	217
6	1.7	.225	48.7	0	8	8.5	83
7	1.8	.175	45.0	0	110	11	230
8	1.8	.200	48.7	.001	8	7	138
9	1.8	.225	41.3	.002	59	8.5	158
10	1.6	.175	48.7	.002	59	7	238
11	1.6	.200	41.3	0	110	8.5	189
12	1.6	.225	45.0	.001	8	11	99
13	1.7	.175	45.0	.002	8	8.5	123
14	1.7	.200	48.7	0	59	11	198
15	1.7	.225	41.3	.001	110	7	205
16	1.8	.175	48.7	.001	110	8.5	115
17	1.8	.200	41.3	.002	8	11	202
18	1.8	.225	45.0	0	59	7	208

Table 7-2 to guide us in the generation of the data using the robotics model. In an experiment, we would have had to make physical runs of the combinations of inertia, motor load, spring load, torque, and part weights and sizes. The responses for this first noise matrix are found under the response column in Table 7-2. We now will set up the other eight noise matrices and obtain data in a similar manner. The 18 observations from each noise matrix are summarized in two statistics, the average and the signal to noise (S/N). Table 7-3 contains this information.

Each row of the data matrix (Table 7-3) represents a run from the design matrix and each column represents a run from the noise matrix.

Now we are ready for the analysis of the average and S/N data. We want to maximize the S/N and minimize the average. We will utilize the analysis of variance (ANOVA) method to find the levels of our controlling factors that will accomplish our goals. From these calculations and the plots, we see that the biggest S/Ns and the smallest averages are obtained with low inertia (1.7), at a level of motor load of .275, high spring load (135), and high torque (.03).

Although we have obtained the required information to design the robot, the problem we have encountered is the vast amount of work required to produce the variation (through the outer arrays) that we want to

Table 7-3a. Average Time of Delivery.

Row (from design matrix)	Columns (from noise matrices) 1	2	3	4	5	6	7	8	9	10	11	12	13	14	15	16	17	18	Average time
1																			174.6
2																			112.1
3																			62.3
4																			104.1
5																			141.9
6																			122.1
7																			174.7
8																			123.2
9																			193.4

Table 7-3b. Signal to Noise.

Row (from design matrix)	Columns (from noise matrices) 1	2	3	4	5	6	7	8	9	10	11	12	13	14	15	16	17	18	S/N(S)
1																			-45.11
2																			-41.28
3																			-36.53
4																			-40.71
5																			-43.28
6																			-41.92
7																			-44.96
8																			-42.11
9																			-45.82

minimize. In minimizing variation we have maximized engineering effort. While we may argue that such an investment is well the effort, we should also realize that if there are simpler and less costly methods to achieve the same goal, we should utilize them.

While there are a number of variations (see Appendix 7-B) on the theme of inducing variation in a parameter design, the method that accomplishes the task with the least effort is called the *internal stress method.* The principle behind this method is exactly the same concept behind the inclusion of outer noise factors in the outer arrays. However, in this method, we consider all other factors as "outer noise" acting upon the factor under study.

Table 7-4. Analysis of Average Time.

ANOVA Robot Average Response

SOURCE		SUM OF SQ.	DF	% CONT.	AVERAGE EFFECTS LEVEL-1	LEVEL-2	LEVEL-3
Inert	-(L)	3373.82800	1	24.70	116.3333	122.6667	163.7593
	-(Q)	604.10140	1	4.42			
Mtr Ld	-(L)	950.04340	1	6.96	151.0926	125.7407	125.9259
	-(Q)	326.07200	1	2.39			
Sprg Ld	-(L)	279.40830	1	2.05	139.9445	136.5185	126.2963
	-(Q)	23.09478	1	0.17			
Torque	-(L)	8082.97500	1	59.18	169.9445	136.2778	96.5370
	-(Q)	18.44728	1	0.14			
RESIDUAL		0.00000	0	0.00			
TOTAL		13657.97000	8	100.00			

Table 7-5. Anaysis of Signal to Noise.

ANOVA Robot Signal To Noise Type S

SOURCE		SUM OF SQ.	DF	% CONT.	AVERAGE EFFECTS LEVEL-1	LEVEL-2	LEVEL-3
Inert	-(L)	16.53031	1	25.34	-40.9826	-41.9771	-44.3023
	-(Q)	0.88553	1	1.36			
Mtr Ld	-(L)	7.06485	1	10.83	-43.6008	-42.2307	-41.4305
	-(Q)	0.16235	1	0.25			
Sprg Ld	-(L)	3.16607	1	4.85	-43.0514	-42.6120	-41.5986
	-(Q)	0.16472	1	0.25			
Torque	-(L)	36.83711	1	56.47	-44.7455	-42.7266	-39.7899
	-(Q)	0.42126	1	0.65			
RESIDUAL		-0.00001	0	-0.00			
TOTAL		65.23220	8	100.00			

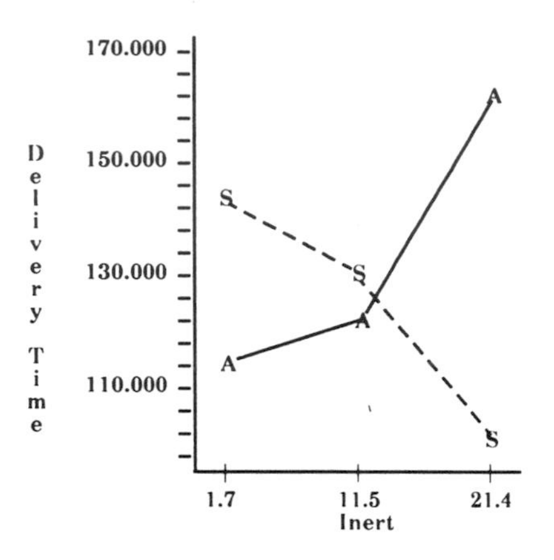

Figure 7-3.

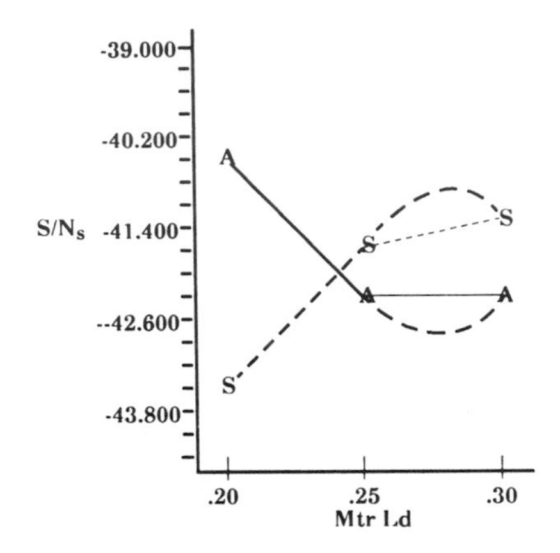

Figure 7-4.

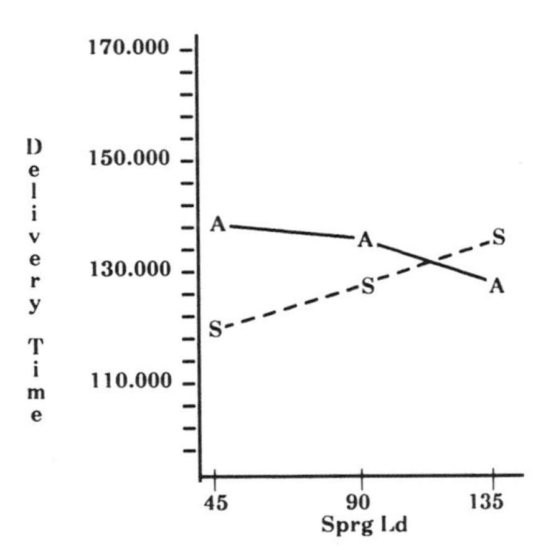

Figure 7-5.

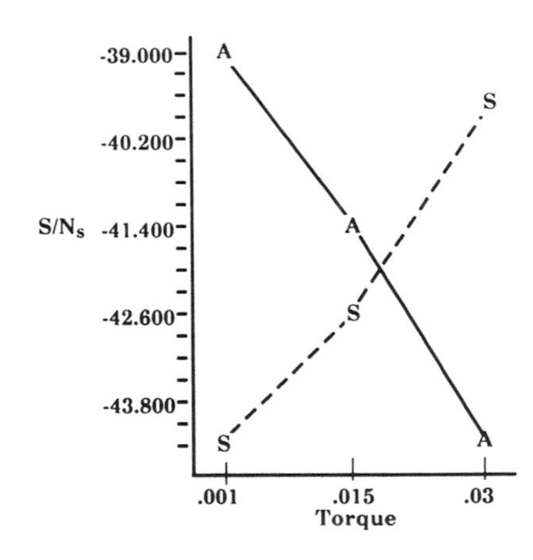

Figure 7-6.

We will now look at the method that allows us to find the robust levels of each factor when all other factors are acting as noise. The basic flow of the experimental investigation still holds for this method. We will design an experiment, gather the data, and then apply a special analysis.

We begin by building the experimental design. We will again use the robotics problem, and therefore will be able to compare the results from this new method with the more costly inner/outer method covered previously. All six factors (including the noise factors) will be placed in the experimental design. We will be forced to use the L27 orthogonal array (OA), since the L9 that we used before can hold only four factors and the L18 has difficulties with the confounding of interactions.

We obtain the responses for this illustrative example from the robotics equation as before. The delivery time for each run of the experiment is shown in Table 7-6. It is important to understand that the levels for each factor are held as closely as possible to the value stated in the experimental configuration. This method does require exacting experimental work.

The data from Table 7-6 is treated using the following concept. Each factor has three levels. It is easy to observe the three levels of inertia (Inert) grouped in the first nine rows (1.7), the second nine rows (11.5), and the last nine rows (21.4). The responses within each group of nine observations are all different. This difference in the response exists because the other factors are changing. This change among the other factors will induce the variation we need to find the robust level of each factor, the level that shows adherance to the target as well as low variation. We compute the average of the responses for each level of the factor. This tells us where that factor optimizes for the target. We also compute the S/N for each level of the factor under study. This tells us about both the location with respect to the target, as well as the robustness of that level to variation in the other factors.

We plot the average delivery times as well as the S/Ns and pick the levels of the factors that have the shortest delivery times and the highest S/Ns (nearest to zero since these are negative S/Ns).

The results of the ISM show the following recommended conditions:

Inertia	1.7
Motor load	0.275
Spring load	135.0
Torque	.03

Table 7-6.

Inert	Mtr Ld	Sprg Ld	Torq	Wght	Size	RESPONSE
1.7	.20	45	.001	8	7	137.8
1.7	.20	90	.015	59	8.5	138.3
1.7	.20	135	.030	110	10	69.7
1.7	.25	45	.015	59	10	132.8
1.7	.25	90	.030	110	7	72.1
1.7	.25	135	.001	8	8.5	55.8
1.7	.30	45	.030	110	8.5	37.6
1.7	.30	90	.001	8	10	82.4
1.7	.30	135	.015	59	7	141.8
11.5	.20	45	.015	110	8.5	122.3
11.5	.20	90	.030	8	10	102.1
11.5	.20	135	.001	59	7	213.0
11.5	.25	45	.030	8	7	101.2
11.5	.25	90	.001	59	8.5	153.2
11.5	.25	135	.015	110	10	98.1
11.5	.30	45	.001	59	10	176.4
11.5	.30	90	.015	110	7	129.2
11.5	.30	135	.030	8	8.5	47.9
21.4	.20	45	.030	59	10	172.1
21.4	.20	90	.001	110	7	243.0
21.4	.20	135	.015	8	8.5	144.0
21.4	.25	45	.001	110	8.5	179.8
21.4	.25	90	.015	8	10	141.9
21.4	.25	135	.030	59	7	152.5
21.4	.30	45	.015	8	7	169.6
21.4	.30	90	.030	59	8.5	121.3
21.4	.30	135	.001	110	10	184.3

In this particular example, we illustrate another advantage of the ISM. We obtain information on the outside noise factors, since they were a part of the experiment. In this case, we cannot pick the best (lowest deliver times), but by using the following reasoning, we can cut the amount of experimentation in the verification/tolerance stage. The reasoning is based on the solid engineering concept of stress testing. From Table 7-7, we observe that the worst delivery times for the noise factors of weight and size are at the midweight and the small sizes. If we can get the robot device to deliver within the 115-ms limit with these stressed conditions, then the device will work at all other conditions. With this type of knowledge, we can cut the tolerance testing considerably, since we do not need to test all noise levels.

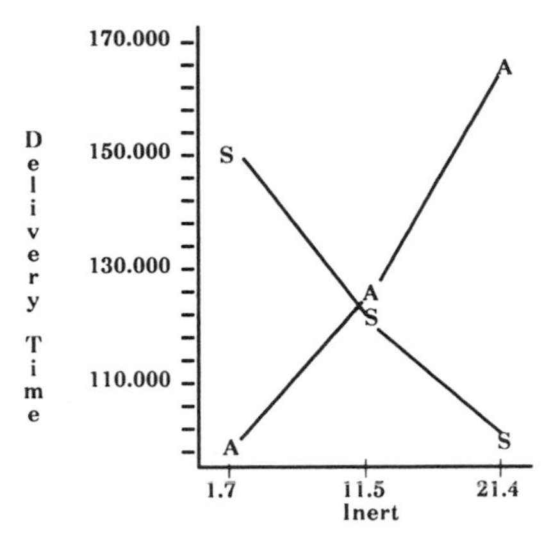

Figure 7-7.

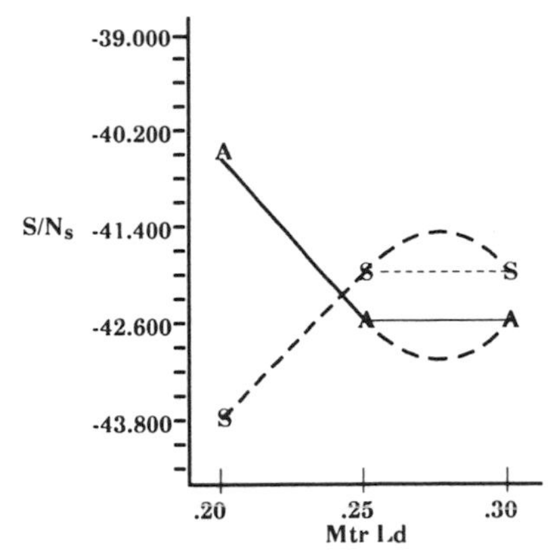

Figure 7-8.

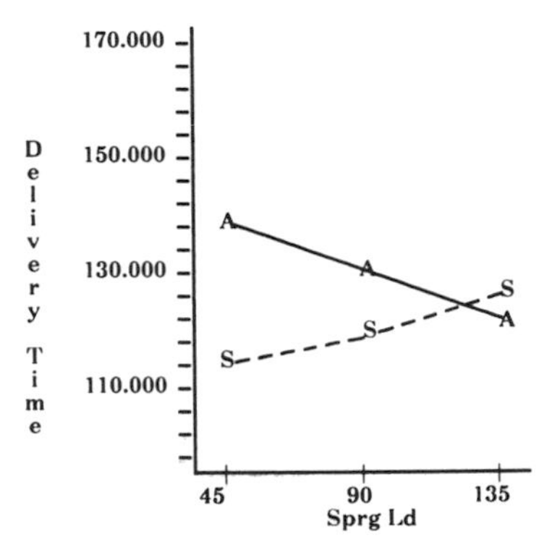

Figure 7-9.

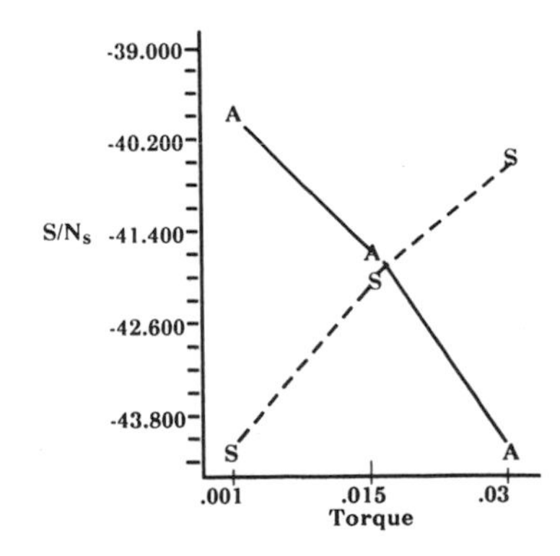

Figure 7-10.

Table 7-7.

Robot Internal Stress Analysis.

Inertia	AVERAGES	SIGNAL TO NOISE (type s)
LEVEL 1	96.47	-40.33
LEVEL 2	127.04	-42.61
LEVEL 3	167.60	-44.65

Motor Load	AVERAGES	SIGNAL TO NOISE (type s)
LEVEL 1	149.14	-43.94
LEVEL 2	120.82	-42.07
LEVEL 3	121.16	-42.38

Spring Load	AVERAGES	SIGNAL TO NOISE (type s)
LEVEL 1	136.62	-43.13
LEVEL 2	131.50	-42.90
LEVEL 3	123.01	-42.58

Torque	AVERAGES	SIGNAL TO NOISE (type s)
LEVEL 1	158.41	-44.50
LEVEL 2	135.33	-42.704
LEVEL 3	97.38	-40.54

Outside Noise Factors

Weight	AVERAGES	Size	AVERAGES
LEVEL 1	109.18	LEVEL 1	151.13
LEVEL 2	155.71	LEVEL 2	111.13
LEVEL 3	126.23	LEVEL 3	128.86

Stress conditions (worst times) for:

Weight	59 oz.
Size	7 in.

The interpretation of these results leads to the same conclusions that we obtained with the more lengthy inner/outer array method. yet the ISM used only 17% of the resources that the inner/outer array method required. A verification/tolerance design would now be applied, as we did in Chapter 6.

WHY THE INTERNAL STRESS METHOD *WORKS*

The ISM works because the robustification, based on the minimum transmitted variation, looks at nonadditive effects, such as the interactions and quadratic effects described at the start of this chapter. Unlike the inner/outer array method, which causes the variation due to these nonadditive effects to take place outside the design matrix, the ISM allows the variation of all the factors to act on the factor being optimized. If there are interactions between this factor and the other factors, or if this factor has a quadratic effect (essentially interacting with itself), then the robust, or least variation, level is identified by the ISM.

In Figure 7-1 we saw how an outer noise factor could interact with a controlling factor. With the ISM, we extend this concept to allow all factors to act as if they were noise. Should the interaction as shown in Figure 7-11 happen in the process we are robustifying, then we choose the level of controlling factor A that shows the least sensitivity to changes in controlling factor B. This means that factor B may vary over a wide range (e.g., have a low- cost, wide tolerance) and not harm the output variation of our process. This means that factor A must be held rather tightly to its "robust" level. What if this is costly? Here is where we may make an engineering trade- off. The interaction may be replotted to show the same information, but this time we will put factor A on the x axis while we draw the isoplot for the levels of factor B. Figure 7-12 shows this configuration. Now we pick the robust level of factor B (which is the low level). This may also be a more economical condition to work with.

Of course, if we include outside noise factors in the ISM, we will pick the level of the controlling factor that withstands the variation in the noise factor best.

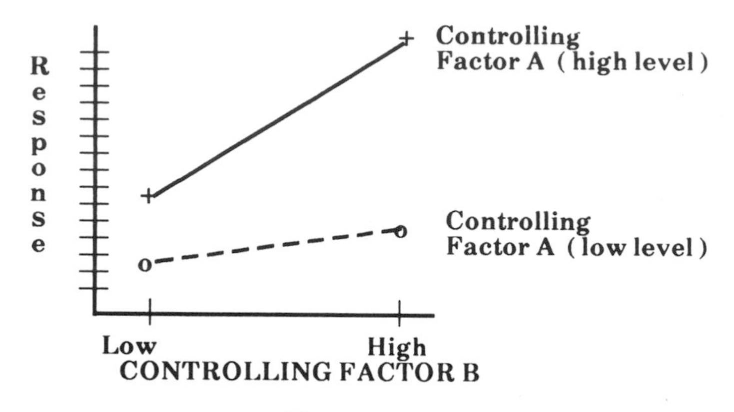

Figure 7-11.

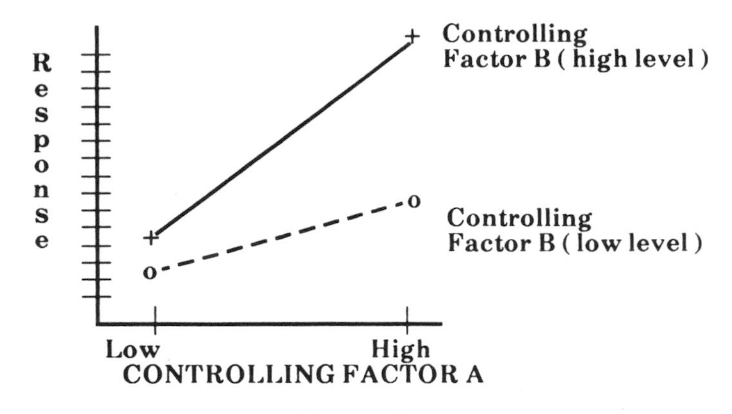

Figure 7-12.

The concept behind the ISM is exactly the same concept behind the more complex and costly inner/outer array method. We will find the robust level—the level of the factor that withstands or is immune to variation. Nonadditive functional relationships form the infrastructure of this robustification process. The structures of experimental designs, as we studied in Chapter 4, allow us to uncover these important functional relationships in an efficient manner.

CAUTIONS AND REQUIREMENTS FOR THE INTERNAL STRESS METHOD

As with all procedures related to the quality infusion process, there must be some rules and warnings to make the method work all the time. The method of inducing variation relies on the fact that each level of each factor is uniformly perturbed by the other factors in the experiment as if these other factors were noise. If the balance of these other factors is different over the levels of the factor under consideration, then the amount of variation induced by the level of the factor depends on the experimental design configuration and not on the "physics" of the problem. This is of course an undesirable result, since if we use another experimental design configuration, we will obtain a different optimum level! The optimum must depend on the functional relationship and not on the experimental design.

Therefore the experimental design configuration must be balanced.

Fortunately almost all experimental design configurations are balanced. A notable exception to this is the central composite design (CCD), which cannot be used in the ISM (although this design [CCD] is ideal for finding equations that describe the process).

The second caution is that there must be at least four active factors in the experiment. If there are fewer than four factors (and these factors must have an influence, e.g., are "active"), the chance for the perturbation that is the key to robustification is diminished. This restriction is not hard to live with, since we often have more factors than we can "shake a stick" at.

The last important concept of the ISM approach to the infusion of quality into an engineering design is that if we have real noise factors, we will include them in the experimental design. This is true since the method relies on the concept that we are treating all factors as if they were noise. A real noise factor included in the experiment can only add credance to the fundamental concept. There is a bonus that arises from the practice of including noise in the experimental design. Besides learning how to robustify the controlling factors, we learn how the noise factors influence the process! This information may be helpful (like it was in the robot example) for the subsequent stages of experimentation, such as the tolerance design, where we must find the influence of the noise factors on the overall variation.

The ISM has been used quite successfully in many applications of the engineering quality by design process. It works because it is built on strong engineering principles and is guided by systematic methods of statistical experimental design.

REFERENCE

Taguchi, G. *Introduction to Quality Engineering.* Asian Productivity Organization, Tokyo, 1986, Chapter 8, Discussion Section.

PROBLEMS FOR CHAPTER 7

1. Using the ISM propose a parameter design for the following ink formulation used in an ink-jet computer printer. Write a statement of intent to be used in obtaining resources to fund this experiment. The goal of the effort is to reduce the blur in the image made by the ink-jet printer by reducing the number of free particles of pigment (carbon black) that surround the ink droplets. Do you think your request for resources will be approved using this approach? Note that this is the same as problem #1 in Chapter 5.

Factors	Levels	Low-cost variation
Surface tension	20, 30, 40 dyn/cm^2	SD = 3 dyn/cm.^{2}
Viscosity	5, 8, 11 centipoise	SD = 1 centipoise
Pigment particle size	2, 4, 6 μm	SD = .5 μm
Number of particles	10, 20, 30 per μm	SD = 1 particle
Printer drive power	.5, .75, 1.0 A	SD = .05 A
Storage	1, 13, 26 weeks	Outside noise

Storage of ink before use is an outside noise factor that could range between 1 and 26 weeks.

2. The breakdown voltage of a semiconductor device has been shown to be a function of the following factors.
   a) Propose a parameter design to find the engineering design configuration to maximize breakdown voltage and to reduce the variation in this response variable. Use the ISM.

b) Write a paragraph or two explaining why the ISM approach will accomplish the goal of the parameter design without as much experimental effort as the design matrix/noise matrix approach that was used in Chapter 5.
c) Do the "low-cost variations" enter into the ISM approach? Why?
d) Where do the "low-cost variations" enter into the engineering quality by design approach?

Factors	Levels	Low-cost variation
Length	0, 1, 2 cm	±20%
Height	−2, 0, +2 cm	±20%
Width	0, 2, 4 cm	±20%
Temperature	160, 170, 180°C	±10%
Speed of rotation	1, 3, 5 rpm	± 5%
Raw material purity	1, 3, 5 ppm	No control, but less than 5 ppm when in production.
Moisture in atmosphere		No control during production and it will change from desert dryness (20% relative humidity [RH]) to 85% RH.

3. I would like to increase the number of miles per gallon (mpg) of gasoline in my automobile by improving my driving habits. Given the following possible factors that could relate to this opportunity:

a) Devise a parameter design using the ISM that would accomplish this goal.
b) Think of two more factors that could influence the mpg and include them in your design.
c) How long do you think such an experiment would take?
d) When would such an experiment be best planned to start during the year?
e) Compared with the answer from Chapter 5, which plan do you think I will implement?

Factors	Possible levels	Low-cost variation
Speed	40–60 mph	5 mph
Fuel octane	87–92	1
Air conditioner	off-on-economy	none
Outside temperature	32–90°F	3°F
Outside humidity	30–90% RH	5% RH

4. The charge voltage in an electrostatic copier depends on the position of the charging wire. The quality of the copy depends on the charge voltage. If the charge is too low, the image will be light. If the charge is too high, the image will be very black, but there will be an accompanying high level of background, which is a gray mist all over the copy. The mechanical engineer must position the bracket that holds the wire to impart the correct charge consistently. Devise a parameter design utilizing the ISM to accomplish this goal.

Factors	Working range	Low-cost variation
Bend angle of the bracket	97–91°	.5°
Thickness of metal	1.9–2.2 mm	.05 mm
Hole locations (from reference points)		
Unthreaded	14.5–16.0 mm	.05 mm
Threaded	14.0–16.0 mm	.05 mm
Torque on fasteners	90–110 in. oz.	2 in. oz.
Outside noise factors		
Relative humidity	30–70% RH	
Current output of power supply	20 μA	.5 μA

APPENDIX 7-A

Robot Function

BACKGROUND. The six factors as well as their levels and the equation used to generate the "data" in this chapter are all based on actual experimental work. However, to protect the actual functional relationships, the names of the factors have been changed. A central composite experimental design structure was used to produce the original data and multiple linear regression generated the equation from that data. The device that was engineered from this investigation was put under continuous test (24 hr per day, 7 days per week). The test was terminated without failure after 7 months.

EQUATION:

Delivery time (ms) = 1470.63 + 0.17 (inertia^2) − 3144. 7 (motor load)
+ 5729.47 (motor load^2) − 0.00084 (spring load^2) − 735.1 (torque)
+ 264 (part weight) − .0146113 (part weight^2) − 225.7 (part size)
+ 12.84 (part size^2) − 0.0245 (part weight)(inertia)
− 30.4 (part weight)(torque).

Table 7A-1.

FACTORS	Working Range	Low-Cost Variation expressed in Std. Dev.[2]
Inertia	1.7 to 21.4 oz./ft.	5%
Motor Load	0.20 to 0.30 amps	10%
Spring Load	45 to 135 oz.	3 oz.
Torque	0.001 to 0.03 ft. lbs.	0.001 ft. lbs.
Part Weight	8 to 110 oz.	5%
Part Size	7 to 10 inches	0.5%

Response: Delivery time in milliseconds (ms).

Specification: delivery time less than 115 ms.

APPENDIX 7-B

Other Approaches to Variation Induction

In this chapter we introduced the ISM approach to parameter design. As we explained, this method is a very important way of accomplishing the goals of engineering quality by design, while doing the job in an efficient manner. We also saw how the nonlinear effects (either quadratics or interactions) provided the foundation for the ISM.

There are other approaches to the study of robustification and within these approaches, there are many variations on the theme. The ISM happens to be one of the variations, but so important a variation that it has become a method in itself. We will now look briefly at two other approaches whereby we induce variation during experimentation with the sole purpose of understanding the causes of variation.

Stress Method

Both manufacturing tolerances (inner noise) and field use variations (outer noise) as well as raw material fluctuations will cause our products to exhibit variation. The job of parameter design is to find the settings of the factors under our design authority that withstand (be robust) to such noises. The parameter design utilizing the inner/outer array method does this very well. But it takes too much work! Another approach to accomplishing our goal of robustification is to take the worst of the noise conditions and simultaneously impose such a stress on our design authority factors. To obtain a contrast and provide a second "sample" from which to compute an expected loss or S/N, we will run our design authority factors under the best settings of the noises. The concept is similar to the inner/outer array method, but with only two runs in the noise matrix (outer array). This method relies heavily on prior engineering knowledge to identify the sources of noise and their levels that will produce the variation. Taguchi favors this approach as evidenced by his recommendations during specific consultations on particular robust design activities. An example illustrating this method may be found in Chapter 9.

Random Sampling Method

A technique often misused to induce the variation for parameter design is based on the idea that randomly repeated replicates of the runs in our design matrix (inner array) will display the same level of variation that the noise matrix (outer array) produces. The motivation is of course to eliminate the tedious effort in setting the conditions in the noise matrix. In theory the method should work if enough random samples are taken. However, in practice, the experimenters often do not take enough samples and worse than that, the samples are not random replicates, but only duplicates. Research (Ref. 1) at the Center for Quality and Applied Statistics at R.I.T. has shown that 24 replicates are necessary to stabilize this method's results. Taguchi does not recommend this approach and calls it "nonsense" (Ref. 2).

Appendix References

1. Holden, A. Determination of the Random Sample Size Required to Replace a Noise Matrix in a Parameter Design. *Transactions of QED 88.* Rochester Institute of Technology, Rochester, NY, 1988.
2. Private communication, G. Taguchi to T. Barker.

8
Special Topics

No book on methods developed by Dr. Taguchi could ever be complete, since he is continuously refining and, more importantly, adding to the already large base of his concepts and techniques. As a world-class engineer, he develops new methods to meet new engineering challenges as the need arises in the many companies for which he consults. In this chapter, we will explore some of these special areas that Dr. Taguchi has developed over the past 40 years.

TRANSFACTOR DESIGN

In some experimental situations where we are investigating different system designs, there are certain common factors between the systems, but there is a factor or factors unique to each system under investigation. For example, in the investigation of methods to harden metal, we could propose using a water quench or an oil-bath quench. The types of steel as well as the annealing temperatures would be common to the experiment, but the water bath would have different temperatures than the oil bath. The oil bath would also have a viscosity factor that the water-bath method did not have.

There are many situations that require a different approach to the experimental configuration to save redundant effort. One such approach is called the *transfactor design.* (See the problems for details on such examples.)

In the following example, we will learn the structure of the transfactor design, how to analyze it, and any drawbacks that will lead to cautions in its application.

Our problem involves the manufacture of resistors (for electronics application). We need to insulate these resistors, and there are two systems that are candidates for this job. Besides the cost of the insulating materials, we need to be sure that the system we choose does its job and prevents electrical flow.

Traditionally, we would set up two experiments to investigate each of the systems. Since we would be insulating similar resistors made in the same way, there is extra effort expended in the two-experiment approach. Table 8-1 shows the factors and their levels for the engineering effort described above.

Table 8-2 shows the experiment for the Bakelite system. There are eight runs for this experiment, which is based on the L8 design found in

Table 8-1.

FACTORS	LEVELS	
Ceramic Types	Type 1	Type 2
Vaporization Material	A	B
SYSTEM: Bakelite Insulation		
Molding Temperature	low	high
SYSTEM: Enamel Insulation		
Enamel types	I	II
Coating times	short	long

Table 8-2. Bakelite Design.

col.# from L8:4		2	1			
Run #	Temperature	Ceramic Material	Vaporization Material	resistance (K ohms) at 5 fixed positions	$\bar{X}$	S/N_T
1	low	1	A	7 10 23 184 206	86.0	-0.1
2	high	1	A	26 28 30 177 190	90.2	1.3
3	low	2	A	22 30 38 118 195	74.0	2.2
4	high	2	A	26 30 45 148 175	84.8	2.7
5	low	1	B	25 31 101 147 165	93.9	3.8
6	high	1	B	46 54 73 153 164	98.0	5.0
7	low	2	B	52 80 96 100 115	88.6	11.3
8	high	2	B	73 89 94 99 108	92.6	16.2

CONTRAST ANALYSIS for Average S/N:	TEMPERATURE	CERMAT	VAPORMAT
	low 4.30	1 2.5	A 1.5
	high 6.30*	2 8.1*	B 9.1*

Chapter 10. Table 8-3 shows the experimental configuration for the enamel system. Again we utilize the L8, but this time with four factors as a half-fractional factorial.

The two separate experiments require 16 total runs. We have essentially looked at two of the factors twice in the two separate experimental configurations. If the behavior of the resistor-forming factors (ceramic types and vaporization material) is the same for both insulating systems (there is *no* interaction) then we have spent too much time and effort in this investigation. For the sake of discovery and understanding, we have completed the experiments in Tables 8-2 and 8-3, and have made the resistance measurements for all 16 runs.

The contrast analysis found in Table 8-2 shows the difference between the average S/N at each of the two levels for the three factors in the Bakelite system design. The high level of the temperature, type 2 ceramic material, and B type vaporization material are all superior levels of the factors if we should decide to use the Bakelite system for insulating our resistors.

The contrast analysts found in Table 8-3 shows the difference between the average S/N at each of the two levels for the four factors in the enamel system design. The type 2 ceramic material and the B type vaporization material are the superior levels of the resistor fabrication process. Enamel

Table 8-3. Enamel Design.

Run #	Ceramic Material	Vaporization Mateial	Enamel Type	Coating Time	resistance (K ohms) at 5 fixed positions	$\bar{X}$	S/N_T
col.# from L8:	4	2	1	7			
1	1	A	I	short	19 24 36 209 220	101.6	-0.3
2	2	A	I	long	20 39 104 155 175	98.6	3.3
3	1	B	I	long	60 94 112 136 157	111.8	8.0
4	2	B	I	short	78 97 112 115 119	104.2	15.1
5	1	A	II	long	70 76 81 105 155	97.4	9.1
6	2	A	II	short	69 77 91 111 115	92.6	13.2
7	1	B	II	short	92 98 106 115 118	105.8	17.9
8	2	B	II	long	93 95 99 104 109	100.0	23.7

CONTRAST ANALYSIS for Average S/N:	CERMAT	VAPORMAT	ENAMLTYP	COATING TIME
	1 8.7	A 6.3	I 6.5	short 11.5
	2 13.8*	B 16.2*	II 16.0*	long 11.0

type II with the greater average S/N should be used, while either coating time is satisfactory if we decide to use the enamel system for unsulating our resistors.

We now know the best settings for the resistor fabrication and the insulation operation. However, which insulation system is superior? We need to decide on only one method, since we cannot afford to build parallel processes in our factory. There are always cost considerations for capital equipment, tooling, etc. What about the quality considerations? The costs of equipment are spent only once. If we have a quality problem, such as too much variation between resistors, we have an ongoing cost that accumulates over the life of the product. In addition, if our competition comes out with a superior (more robust) product, we lose the market share completely!

In this particular situation, the cost and the quality are in step together. The enamel coating system gives resistances that are closer to the target of 100-K ohms with less variation. The enamel process is also less costly from an investment vantage point, since we do not have to purchase large ovens to cure the Bakelite insulator or to buy more costly raw materials!

We have an answer to our system design question, but did we do our engineering job in a quality manner? We ran 16 trials in two separate experiments. We also tested two of the factors redundantly. We have not been efficient.

Another approach to solving this same problem is the *transfactor design.* The name comes from the fact that we run one experimental configuration over (trans) a common set of factors. The common factors (that were redundant in our first experiments) are the ceramic type and the vaporization material.

To accomplish our goal of finding the right system as well as the right system's proper settings *and* the common factors' correct (robust) settings, we will utilize the L8 orthogonal array from Chapter 10. Let's look at this array as shown in Table 8-4.

Since we have two systems to investigate, it is convenient to use column 1 to see the separation of the two systems, although any other column could have been used to do so. We place the enamel system in the top half of column 1 and the Bakelite system in the bottom half of column 1. There are two types of enamel that must be tested, and there are two tem-

Table 8-4.

$L8\ (2^7)$	STRATEGY TABLE							"WESTERN EQUIVALENT"						
Column number:	1	2	3	4	5	6	7	1	2	3	4	5	6	7
Run Number	o	o	x	o	x	x	o	C	B	-BC	A	-AC	-AB	ABC
1	1	1	1	1	1	1	1	-	-	-	-	-	-	-
2	1	1	1	2	2	2	2	-	-	-	+	+	+	+
3	1	2	2	1	1	2	2	-	+	+	-	-	+	+
4	1	2	2	2	2	1	1	-	+	+	+	+	-	-
5	2	I	2	1	2	1	2	+	-	+	-	+	-	+
6	2	1	2	2	1	2	1	+	-	+	+	-	+	-
7	2	2	1	1	2	2	1	+	+	-	-	+	+	-
8	2	2	1	2	1	1	2	+	+	-	+	-	-	+

LINEAR GRAPH FOR L8

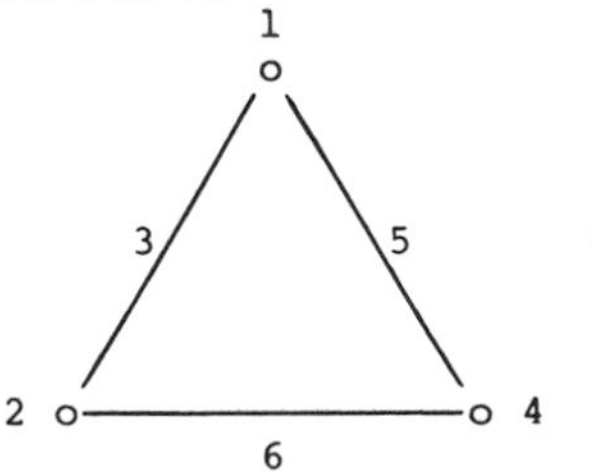

NOTE: o Means OK to use column
x Means do not use, involved with interactions from other columns.

Defining Contrast: 1=ABCD

peratures under investigation in the Bakelite process. Column 2 provides a convenient position for these factors, but note that the enamel type only applies to the enamel process, so it only shows up in the top four rows of column 2. Similarly, the temperature factor applies only to the Bakelite process, so it is positioned in the four bottom rows of column 2.

Let us think for a moment. We have "created" an interaction! In going across (transfactor) our two systems, and by looking at conditions unique to those two systems, we do not have the assurance of physical additivity of the effects in these two (1 and 2) columns. If we were to analyze this interaction, we would need to look to column 3 (as the linear graph states) for the interaction effect. Since the interaction does not have a meaningful interpretation, we will ignore it, but we may not place another factor in column 3, or else the interaction will combine with that factor and give a biased, incorrect result. So we skip column 3 and go to assign the next item to column 4, but this item (the coating time) applies only to the enamel system. Since the enamel system appears only in the first four runs of the experiment, we use only half of column 4 to assign the short and long levels of coating time. We leave the bottom four rows of column 4 empty.

Once again we have "created" an interaction, this time due to a lack of physical balance in the experiment. The interaction is with the factor in the first column. Once again, the linear graph shows that the interaction between column 1 and column 4 falls in column 5. We will not use column 5 for any factor assignments to prevent the production of erroneous information.

There are only two columns left and we have two factors remaining. We assign the ceramic type to column 6 and the vaporization material to column 7. These columns have not been touched by any of our "induced" interactions, but if there are any interactions inherent to the process, they will superimpose themselves on our results and we will never know they are there until we try to reproduce the settings that the experiment recommends. If the predicted results appear, there were no interactions. If the predicted results are not confirmed, then there are probably interactions. Taguchi comments on the later situation with the advice that the system design is not robust enough if there are interactions and we need to search for a new system. Some engineers and most management do not like to hear those words. It might be better to spend an extra bit of effort up front confirming our assumptions about interactions, before we plunge into a highly confounded experiment. Then, however, we may not have the re-

Table 8-5. Transfactor Design.

col.# from L8:	1	2	4	6	7			
Run #	SYSTEM	Enamel/ Temp.	Coating Time	Ceramic Material	Vapor Mateial	resistance (K ohms) at 5 fixed positions	$\bar{X}$	S/N_T
1	ENAMEL	I	short	1	A	19 24 36 209 220	101.6	-0.3
2	ENAMEL	I	long	2	B	98 99 102 110 123	106.4	18.0
3	ENAMEL	II	short	2	B	93 98 100 102 109	100.4	24.6
4	ENAMEL	II	long	1	A	70 76 81 105 155	97.4	9.1
5	BAKELITE	low	e	1	B	25 31 101 147 165	93.8	3.8
6	BAKELITE	low	m	2	A	22 30 38 118 195	80.6	2.2
7	BAKELITE	high	p	2	A	26 30 45 148 175	84.8	2.7
8	BAKELITE	high	t	1	B	46 54 73 153 164	98.0	5.0
			y					

	TEMP	ENAMLTYP	COATING TIME	CERMAT	VAPORMAT	SYSTEM
CONTRAST ANALYSIS for	low 3.0	I 8.9	short 11.2	1 4.4	A 3.4	ENAML 12.9
Average S/N:	high 3.9	II 16.9	long 13.4	2 11.9	B 12.9	BAKEL 3.4

sources to do so. In that case, it is better to approach the problem in a systematic, organized way, rather than in a random, disorganized manner.

"IT IS BETTER TO DO SOMETHING THAT IS SLIGHTLY IMPERFECT THAN TO DO NOTHING PERFECTLY!"

Let's see if our transfactor design gives us any different results from the two-stage approach. Table 8-5 shows the design, data, and analysis. First, we note that the enamel system is superior to the Bakelite system. Given that we decide upon the enamel system, we would then pick the high S/N averages for the enamel factors. We see that the type II enamel is best and that the short coating time is slightly better than the longer coating time. We now have our recommended insulation system and its best conditions.

We now decide on the resistor fabrication factor levels. We see that the type 2 ceramic has the larger S/N and the B vaporization material is superior with a larger S/N.

While the numerical values are not exactly the same when we compare the 16-run approach with the transfactor approach, we see that the engineering recommendation is exactly the same in both cases. Since engineering recommendations are the goal of our experiments, we have done our job. If a scientific understanding of the mechanisms were our goal, then we have not accomplished anything.

The many critics of the Taguchi approach to quality control simply do not understand the goals of the engineering quality by design process. They look upon experimental design as an exact science with only one correct answer. Experimental design in the hands of engineers is a tool to help them find a plausible (not necessarily unique) answer to complex engineering design problems. Such plausible answers are doubly constrained by a need to find both the correct set point as well as the lowest variation around that set point. The power of the S/N is its ability to accomplish this task with a minimum of analysis effort. The raw data in this example illustrates this feature of the Taguchi approach to the quality infusion process.

TRANSFACTOR DESIGN IN GENERAL

Before we leave the subject of the transfactor design, we should explore the basic concepts and show a general approach to the utilization of this concept. We will first look at the engineering considerations and then at the experimental structure requirements.

Table 8-6 shows the four engineering considerations, which include the ideas that we are looking for a system design, we are also looking for

Table 8-6. Engineering Considerations.

1) We are investigating the feasibility of system designs.
2) The systems have different factors with levels within these factors that need to be selected.
3) There is at least one common factor between the systems.
4) The system factors and common factors do NOT interact.

the levels within the factors of this system design, we have common factors, and there is no interaction between any of the factors under investigation.

The selection of the proper orthogonal array (from Chapter 10) depends on the number of factors and the subfactors within the investigation. In our example, we had two systems and two common factors. Within the systems, there were either one or two subfactors. Since the choice of the orthogonal array depends on the most information we need, we add up the number of total factors, including subfactors, to determine the number of columns required to accomplish the investigation. In our case, we needed five columns and each factor was at two levels, so the L8 (with seven columns) was the minimum design configuration. Table 8-7 shows the steps in selecting the proper array.

$$\text{\# columns} = 1 + \text{\# sf} + \text{\# cf} + \text{\# int}, \tag{8-1}$$

where # columns is the number of columns in a two-level OA, # sf is the maximum number of subfactors in the systems, # cf is the number of common factors, and # int is the number of "created" interactions.

Transfactor is a special topic and a special application that can save effort and is especially useful in the selection of the system design. It won't be used in our every-day applications of engineering quality by design, but it is useful to understand its power and limitations.

Table 8-7. Experimental Structure Considerations.

1) Determine the number of candidate system designs. (Two systems will make the experimental effort easier)

2) Determine the number of sub-factors within each system.

3) Determine the number of common factors.

4) The size of the orthogonal (OA) array depends on the number of columns in the array that will be equal to or greater than the sum of the maximum number of sub-factors and common factors plus the column for the system plus any interactions "created" by the configuration.

MIXED LEVELS IN ORTHOGONAL ARRAYS

In the transfactor application of OAs, we saw how the linear graph helps in determining the proper placement of factors to avoid unnecessary confounding of effects that we need to understand with interactions. The linear graphs have uses beyond this important concept. The following is one such application.

The OAs presented in Chapter 10 fall into two categories. There are the two-level arrays and the three- level arrays. What if we have an experimental situation that required four levels, or, more realistically, what if some of the factors were at two levels, some were at three levels, and others were at four levels?

With quantitative factors, we could tailor the problem to fit the available design configuration. A quantitative factor such as temperature will have a working range of 100°-200°C. There is usually no difficulty in dividing this range into three levels (100, 150, 200) or simply into two levels (100, 200). How we pick the levels for such quantitative factors depends upon our expectation of the degree of curvature induced by the factor. If we expect a strong quadratic relationship, we most certainly must use a three-level design such as the L9 or L27.

However, if we expect a simple straight-line, linear relationship between the response and the factor, a two-level design is in order. Quantitative factors are comparatively easy. With category-type factors that include concepts, systems, vendors, or methods, it is not easy or even possible to make a logical subdivision between the category levels. If there are only two vendors (A and B), we are unable to put this factor into a three-level design, since there is no such level of vendor as A− or B+.

Let's look at a more specific example of this problem and explore two different approaches to a solution. We are investigating the relationship between the molecular weight (MW) of a polymer and the temperature and pressure in the reaction. We have two vendors who will supply the monomer for the reaction. Table 8-8 shows these factors and the working ranges.

Since we suspect that there could be a quadratic relationship between MW and the temperature, we would be inclined to use the L9 design for the experiment. However, by using such a three-level structure, we imply that it is necessary to find another vendor to fill the third level of this factor! Table 8-9 shows one approach to accomplishing our goal without finding another level to fill the design structure. Taguchi calls this the *dummy-factor* method. We simply fill the third empty level of the three-

Table 8-8.

FACTOR	WORKING RANGE
	o o
Temperature	350 to 500 F
Pressure	20 psi to 50 psi
Vendor	Monsanto, Du Pont

level design structure with one of the previous levels, as shown in Table 8-9. This action has a deteriorating effect on the balance of the experiment, but is an expedient method that accomplishes our goal.

The choice of the dummy level is usually based on prior engineering information. This information could include such considerations as "this vendor is more likely to be the prime vendor," or "we have had trouble with this vendor before, so look at them with more frequency," etc. If there is no prior information to select one level over another for the dummy level, then a random choice (coin toss) is appropriate.

The problem with the dummy-factor approach is the loss of balance in the overall design. There is another way in which we can combine quantitative and categorical factors and still retain the balance of the experimental structure. Before we do that, however, we need to learn how to expand the number of levels in an OA.

Table 8-9.

col.#	1	2	3	1	2	3	
	VENDOR	TEMP	PRESS	A	B	C	
	Monsanto	350	20	1	1	1	
	Monsanto	425	35	1	2	2	
	Monsanto	500	50	1	3	3	
	Du Pont	350	35	2	1	2	
	Du Pont	425	50	2	2	3	
	Du Pont	500	20	2	3	1	
	Du Pont	350	50	3	1	3	While the third level of
	Du Pont	425	20	3	2	1	factor A should differ,
	Du Pont	500	35	3	3	2	it is kept the same to accommodate only 2 levels.
	L9 Physical Units			L9 Design Units			

EXPANDING THE NUMBER OF LEVELS

In a basic two-level OA, there are pieces of information that will probably show little if any influence on the response. This is especially true if there are only a small number of factors in the experiment. Just as we learned in Chapter 4 that it is possible to reallocate some of this "empty" information with other, more useful information, we will now observe how that same thought process may be applied to expanding the number of levels.

With the basic L8 design found in Chapter 10, it is possible to investigate as many as seven factors, *if* there are no interactions among these factors. From an information perspective, we are able to derive seven independent pieces of information with the eight runs. The information may be allocated to more factors (as was done in Chapter 4), or we may allocate the information to more levels.

We count information with degrees of freedom (df). Each df is "earned" by performing a test run in the design matrix. Each level that we add to a factor requires one more df. It takes two levels to produce the fundamental information unit, which is merely the difference between the low level and the high level. If we add one more level, we may look at the quadratic relationship. A fourth level allows a look at the basic linear (1 df), quadratic (1 df), and cubic (1 df). Each time we add another level, we are able to resolve a higher order polynomial fit to the response. Since most functions confine their relationships to no more than a quadratic, we usually do not need any more than a three-level design to find the required information.

If we were to need a four-level design we could construct it from a two-level base structure if we are careful and apply our prior knowledge of the process under investigation.

Table 8-10 is the basic L8 structure. Using our molecular weight example, we will expand the number of levels of the temperature to four, put pressure at only two levels, since we do not expect a curved relationship with it, and do so in the context of only eight runs. We will begin with temperature. Observe in Table 8-10 that if we look at the sets of conditions in both columns 1 and 2 together, there will be four such unique sets. Table 8-11 shows these sets.

Each set of the combined columns 1 and 2 makes up a new level. Since we have four sets, we now have a four-level factor. However, with four levels we will need three df, as discussed above. Columns 1 and 2 contribute two of these three df. Where does the last df come from? We look to the linear graph and find that column 3 lies on the line between columns 1

Table 8-10.

$L8\ (2^{7})$ STRATEGY TABLE — "WESTERN EQUIVALENT"

Column number:	1	2	3	4	5	6	7	1	2	3	4	5	6	7
Run Number	o	o	x	o	x	x	o	C	B	-BC	A	-AC	-AB	ABC
1	1	1	1	1	1	1	1	-	-	-	-	-	-	-
2	1	1	1	2	2	2	2	-	-	-	+	+	+	+
3	1	2	2	1	1	2	2	-	+	+	-	-	+	+
4	1	2	2	2	2	1	1	-	+	+	+	+	-	-
5	2	1	2	1	2	1	2	+	-	+	-	+	-	+
6	2	1	2	2	1	2	1	+	-	+	+	-	+	-
7	2	2	1	1	2	2	1	+	+	-	-	+	+	-
8	2	2	1	2	1	1	2	+	+	-	+	-	-	+

LINEAR GRAPH FOR L8

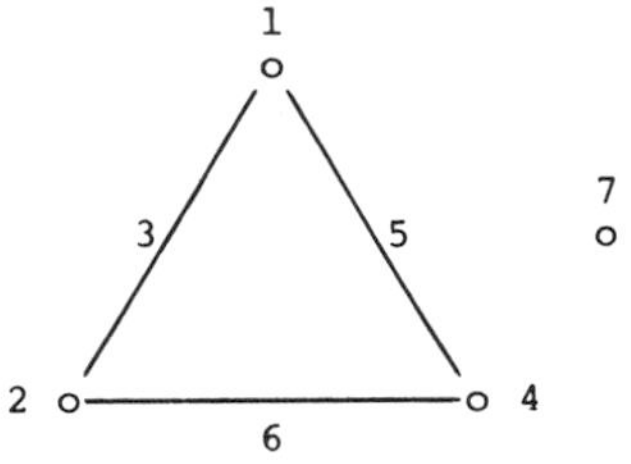

NOTE: o Means OK to use column
x Means do not use, involved with interactions from other columns.

Defining Contrast: 1=ABCD

Table 8-11.

Col:#	1	2		3	EXPANDED 4-level factor
	1	1		1	1
	1	1	Set I	1	1
	1	2		2	2
	1	2	Set II	2	2
	2	1		2	3
	2	1	Set III	2	3
	2	2		1	4
	2	2	Set IV	1	4

and 2. We have placed column 3 in Table 8-11 and, sure enough, it does not change the set structure originated in columns 1 and 2. Column 3 is not independent of columns 1 and 2. (The "Western equivalent" shows that column 3 is actually the mathematical product of columns 1 and 2.) Column 3 provides the final df for our four-level factor.

Now let's put it all together and build an experimental structure with four levels for temperature and two levels for pressure. We will use column 4 for pressure. Table 8-12 shows this design configuration.

There is one last piece of information that we must account for in the above design. If there is an interaction between temperature and pressure, it will require three more df (remember the df in the interaction is the product of the df of the interacting factors). Columns 5, 6, and 7 of the L8 total to those 3 df. Therefore, if there is an interaction between temperature and pressure, we may not place any other factors in the last three columns. However, if we can rule out the interaction, it is possible to utilize any one of these columns to add another factor, such as the vendor factor we put in as a "dummy" in the L9 design. Table 8-13 shows the vendor factor placed in column 7 (although columns 5 or 6 are equally useful).

We have produced a mixed-level design without altering the balance, as we did in the dummy-factor approach. However, there is a lurking danger with the design in Table 8-13. If an interaction exists among these

Table 8-12.

col.:	1&2	4	1&2	4
	A	B	TEMPERATURE	PRESSURE
	1	1	350	20
	1	2	350	50
	2	1	400	20
	2	2	400	50
	3	1	450	20
	3	2	450	50
	4	1	500	20
	4	2	500	50
	Design Units		Physical Units	
	3df	1df		

Table 8-13.

col.:	1&2	4	7	1&2	4	7
	A	B	C	TEMPERATURE	PRESSURE	VENDOR
	1	1	1	350	20	Monsanto
	1	2	2	350	50	Du Pont
	2	1	2	400	20	Du Pont
	2	2	1	400	50	Monsanto
	3	1	2	450	20	Du Pont
	3	2	1	450	50	Monsanto
	4	1	1	500	20	Monsanto
	4	2	2	500	50	Du Pont
	Design Units			Physical Units		
	3df	1df	1df			

factors, that interaction will be confounded with the single effects and we will obtain tainted results. This is the warning that goes with any form of fractional factorial design. However, if we do not utilize such designs, we do not attack the problem systematically. As we have said before, use as much prior knowledge in the construction of the experimental designs and the designs will deliver new knowledge for you.

In this section we have seen how to mix levels with the dummy-factor approach and how to expand the number of levels in a basic OA. There are many variations on this concept waiting for your creativity to capture them. If you follow the basic steps and count df and ask where the confounding will take place, your creative application of this special topic will be well directed.

RANDOMIZATION

There are a number of implications built into the OA structures presented by Taguchi. We have seen and discussed the most obvious of these implications, namely, the apparent license to fill a design with factors. This implied license might lead us to think it is possible to place seven factors in an L8 design. We know better than that and realize that certain requirements must be in place to build such a "saturated" (all df used for single effects) design.

However, there is another implication of the designs in Taguchi's OA catalog—the run order. The statistical experimental design community has always admonished that the tests in an experiment be run in a random order, rather than the systematic order shown in the design catalog. We will explore the reason behind this advice. However, the tables of OAs as presented by Taguchi seem to tell us *not* to randomize!

Look at the L27 design in Chapter 10. Notice that the frequency of change between levels in the first column happens only twice. All the low levels (1s) are grouped together in sets of nine, all the midlevels (2s) are grouped together in sets of nine, and all the high levels (3s) are grouped together in sets of nine. Taguchi calls the factor in this column a *group 1* factor. It is a factor that is difficult to change and therefore does not change with as much frequency as the factors in the remainder of the experimental design structure.

Again, in the L27 we observe that the frequency of change of the factors in the second, third, and fourth columns comes in three sets with three low levels (1s) in a set, three midlevels (2s) in a set, and three high (3s) in a set. These columns contain factors from *group 2,* which are considered easier to change than group 1 factors, but not as easy as a group 3 factor.

The *group 3* factors' frequency of change is most often and these factors are considered easy to change. The group 3 columns make up the remainder of the L27.

The temptation is to simply assign the factors to the experimental structure and to run it in systematic order. This temptation is easy to succumb to if we have such hard-to-change factors, say, in an automobile engine. The compression ratio requires an almost complete disassembly of the engine block, so it is logically a group 1 factor. In fact, if we were to randomize the compression ratio. the cycle of assembly-disassembly would probably do more harm than good, since we probably would break the engine in the process! The fuel rail would be a little easier to attach to the engine, but it still is a time-consuming operation, so it would be likely to be placed in a group 2 column. Changing the spark plugs is a relatively minor and easy task, so it would be placed in a group 3 column.

We will now look at the consequences of running systematically rather than randomly. Tables 8-14 and 8-15 show how randomization prevents incorrect decisions from being made. In Table 8-14 a two-factor experiment involving temperature and pressure is shown. However, another unknown factor is systematically varying along with these two factors. This unknown has a strong negative effect (but we do not know that). In Table 8- 14 the experiment is run in the order of the L9 experimental structure and the results are a function of the factors we are studying as well as

Table 8-14.

Order#& Run#	TEMP	PRESS	level of unknown factor	resp.
1	100	14	1	100
2	100	28	1	100
3	100	42	1	100
4	200	14	2	100
5	200	28	2	100
6	200	42	3	0
7	300	14	3	100
8	300	28	4	0
9	300	42	4	0

NOT Randomized order

the unknown. In Table 8-15, the run order is randomized. There is a vast difference in the results of these two experiments.

The analysis of these two experiments shows the high degree of confusion that can take place if unknown factors change in a systematic manner along with the factors we are trying to understand. Randomization guards against such confusion. (By the way, the true function in this experiment is a positive influence by the temperature. This is correctly illustrated in Figure 8-2, the plot of the results from the randomized run. Compare this true result with the false result found in Figure 8-1.)

However, randomization is a burden on the experimenter. It is often possible that if randomization is forced on an experimenter, and the changing of the factors is an expensive, time-consuming process, the experiment will not get done. Now the experimenter is in a real dilemma. Randomization of the experiment, which will produce good results, takes too long. Systematic order gets the job done, but could lead to distorted results. What to do? In reality, we usually run systematically and use our *engineering* abilities to identify and control the unwanted factors. We remove the burden of randomization from one shoulder and take the burden of extra care and control on the other shoulder. Experiments without

Table 8-15.

Order#	Run#	TEMP	PRESS	level of unknown factor	resp.
1	5	200	28	1	200
2	7	300	14	1	300
3	2	100	28	1	100
4	9	300	42	2	200
5	6	200	42	2	100
6	1	100	14	3	-50
7	8	300	28	3	100
8	3	100	42	4	-200
9	4	200	14	4	-100

Randomized order

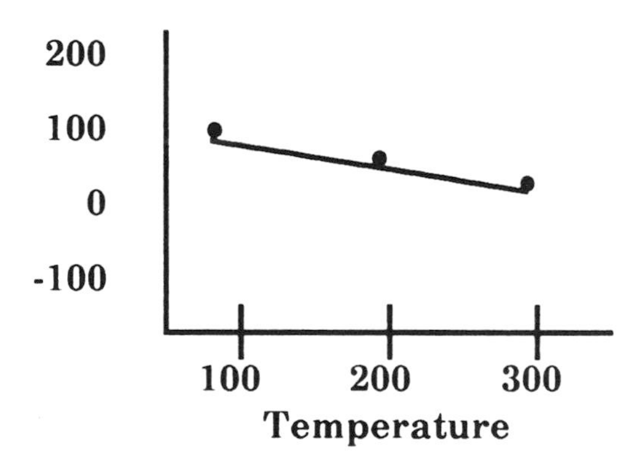

Figure 8-1.

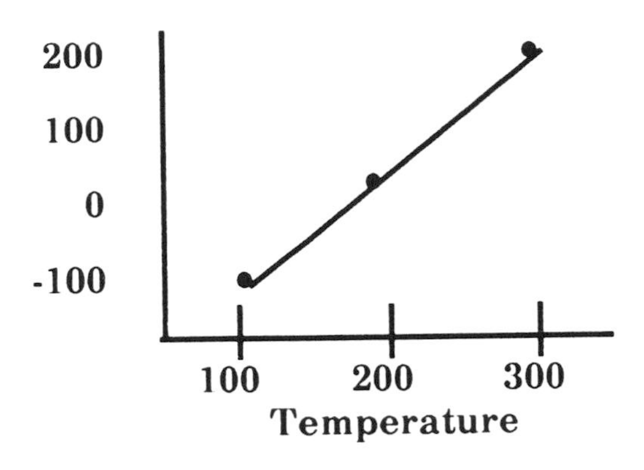

Figure 8-2.

randomization must not be taken lightly, no matter what the OA structures seem to imply.

Taguchi, like so many brilliant people, has a difficult time conveying the entirety of his thoughts to his listeners. Many of the between the lines or intuitively obvious parts of his philosophy and techniques are never stated in his writings, simply because they are too obvious to someone who has practiced the methods for so long. Randomization is a tool that is nonsensical if we use other methods to compensate for the influence of factors beyond the set in our experiment. However, if we do not use the engineering tool of control over factors beyond the set in our experiment, then the only way to prevent bias in the results is to randomize.

"To randomize or not randomize? That is the question " simply boils down to how much we know and how far we can apply this knowledge. The more we know, the less we have to do. The less we know, the more we have to do. Sounds familiar doesn't it? It should, because it is the basic theme of this chapter on special topics.

ATTRIBUTE DATA

All of the discussion in this chapter thus far has centered around the design of the experiment. Designing the experiment correctly is essential. If we get the design right, the analysis will follow easily. However, in some situations, the analysis does become a problem due to the nature of the response variable. In Chapter 4, we mentioned that a response to our experimental conditions should be quantitative, precise, and meaningful. While we seek out such qualities in a response, we sometimes are stuck with a response that simply describes an attribute of a quality characteristic.

A part may fail after a long reliability test. A copy may be good or bad. There may even be a distribution of failures of parts depending on the time of failure. These are realistic problems. Taguchi has developed a method for dealing with categorical data based on attributes. This method is called *accumulation analysis.*

Accumulation analysis is unfortunately, overly complicated (1) and inherently incorrect. Rather than demonstrating the complexity of accumulation analysis by filling the following pages with formulas to compute an ANOVA that is faulty anyway, we will investigate the inherent problem with the method and then suggest better ways to accomplish our goal.

In Chapter 5, Table 5-4, we have image quality (IQ) data that describe on a continuous scale the relative "goodness" of a series of images. Such data could have been placed into categories such as excellent to poor. To illustrate our point, we will classify the numerical data from Table 5-4 using this set of rules:

Taking the eight runs of the IQ design, we will classify the images according to the attributes in Table 8-16. Table 8-17 shows the results of this effort. Remember, we are performing this classification to illustrate a point, not to recommend the practice!

Now for the inherent problem. We take the frequency in each class for each run and find the accumulated (and thus the name of the method) frequency up to that point. Table 8-18 shows the result of this effort. If we utilize these cumulative frequencies categories for our analysis, we have violated the rules of independent information. Each subsequent category in Table 8-18 uses previous information, and from an information theory

Table 8-16.

Classes	Attribute	numerical rating from 5-4
1	Excellent	9-10
2	Very Good	7-8
3	Good	5-6
4	Fair	3-4
5	Poor	1-2

Table 8-17.

run #	frequency of copy in class 1	2	3	4	5
1	0	1	4	3	0
2	0	1	0	6	1
3	4	0	1	3	0
4	4	3	1	0	0
5	0	1	2	3	2
6	0	0	1	3	4
7	4	1	2	1	0
8	8	0	0	0	0

view, we tried to create df. We know that df may not be created. Degrees of freedom must be earned by doing experimental work. The experiment we are looking at (L8) has only 7 df. From the data in Table 8-18, Taguchi's analysis would indicate that we have 4 df for each effect (or four times the actual df). This is untrue, since we have reused the data so many times in our accumulation. We could further fabricate more df by taking more categories!

We will not pursue any further analysis of this accumulation method, but we cannot leave the concept of what to do when there does not seem to be a continuous response.

Table 8-18.

run #	Accumulated frequency categories I	II	III	IV	V
1	0	1	5	8	8
2	0	1	1	7	8
3	4	4	5	8	8
4	4	7	8	8	8
5	0	1	3	6	8
6	0	0	1	4	8
7	4	5	7	8	8
8	8	8	8	8	8

The lasting solution to this problem is to develop a continuous response variable to *measure* the quality characteristic. The field of experimental psychology has developed many such *psychometric* responses. This field is devoted to taking attribute judgments and quantifying them into meaningful, precise measures of the characteristics under study. This subject is beyond the scope of this book, but the references to this chapter can point the interested engineer in the right direction.

REFERENCES

1. Box, G., Jones, S. *An Investigation of the Method of Accumulation Analysis.* Center for Quality and Productivity Improvement, University of Wisconsin-Madison, Report No. 19, December, 1986.
2. Gescheider, G. A. *Psychophysics Method and Theory.* Lawrence Erlbaum Associates, Hillside, NJ, 1976.
3. Guilford, J. P. *Psychometric Methods.* McGraw Hill, New York, 1954.
4. Torgerson, W. S. *Theory and Methods of Scaling.* Wiley, 1958.

PROBLEMS FOR CHAPTER 8

1. Propose a transfactor design for the following steel-hardening experiment:

Table 8-P1.

FACTOR			LEVELS	
Steel type			1040	1080
Annealing Temperature			$1000^{\circ}C$	$1400^{\circ}C$
Quenching Method	Water Bath Temperature		$50^{\circ}C$	$90^{\circ}C$
	Oil Bath	Temperature	$100^{\circ}C$	$150^{\circ}C$
		Viscosity	Thin	Thick

If the transfactor design had not been used, show how much extra effort would be necessary to complete the investigation.

2. Use a transfactor design for the following problem:

Table 8-P2.

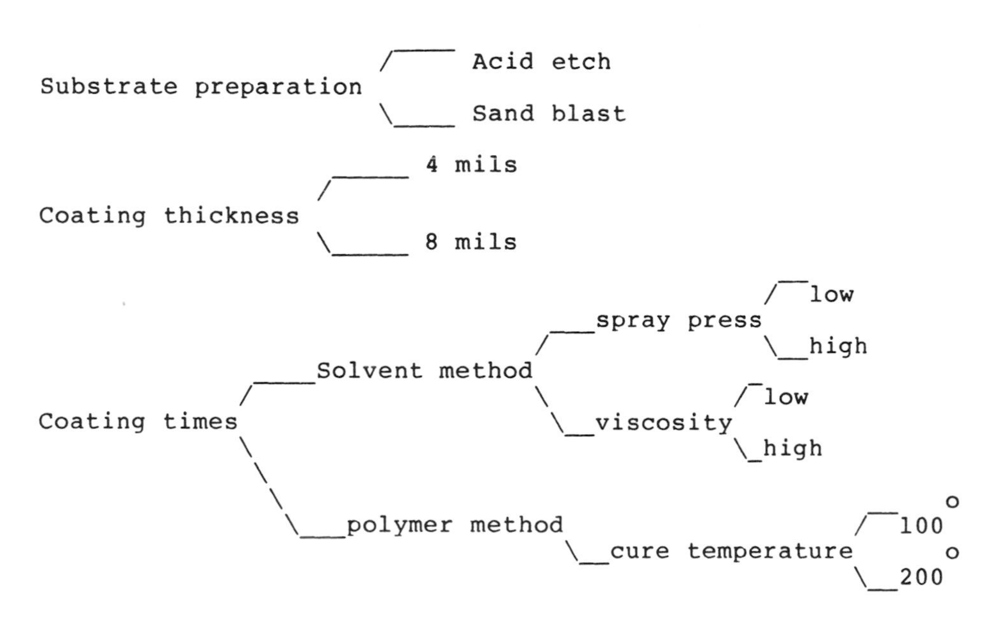

3. Set up an experimental design structure for the following conditions. Do not use more than 16 runs.

	Levels
Temperature	100, 200, 300, 400°
Pressure	15, 30 psi
Time	15, 20, 25 min
Catalyst	.5%, .75%

9
Examples of Engineering Quality by Design in Action

In this chapter we will look at a number of applications of the engineering quality by design process in action. These applications cover a wide range of disciplines and should help acquaint the reader with the engineering approach to using the tools we have outlined in this book.

PHOTOGRAPHIC DEVELOPER

We will begin our examples with a chemical engineering problem taken from the field of photographic science and engineering. Our company makes and distributes the developer powder that is mixed with water to make the developer solution that changes the latent image on the film into a visible image. We also are able to recommend the processing conditions that will work best with our developer formulation.

While our developer has worked in the past to produce adequate quality images, competition (especially from Japan) has raised the level of expectation of our customers and we are rapidly losing sales. Our marketing staff estimates that we have one year to improve the developer performance, before we are out of business in that product line completely.

Table 9-1.

FACTOR NAME	Working Range	Variation (1 standard deviation)
Metol	5 to 8 g/l	.25 g/l
Hydroquinone	13 to 16 g/l	.25 g/l
Sodium Sulfite	40 to 80 g/l	1.0 g/l
Alkali	40 to 80 g/l	1.0 g/l
KBr	1 to 5 g/l	.2 g/l
Temperature	70 to 90 ^{o}F	.5 ^{o}F
Speed of film travel	5 to 8 inches/sec.	.1 inches/sec.

An interaction is known to exist between the Metol and the Hydroquinone.

In the brainstorming meeting and subsequent follow-up to that meeting, the developer team identified the following factors and the working range, as well as the possible interaction. They also know from current production SPC data the variation expected for each of the factors. This information is found in Table 9-1.

Since this is a situation that does not involve any outside noise factors and since it would be very difficult as well as costly to conduct this experiment using the inner/outer array approach, the team decides to utilize the internal stress approach. Also, since there are possible curvilinear effects, a three-level experimental structure is appropriate. The L27 found in Chapter 10 is the structure for this experiment. Table 9-2 shows this design with the actual factors and their levels, and the 27 responses obtained after the experiment was run.

As we look through the responses in Table 9-2, we observe that one of the runs produced an image-quality (IQ) value of 100. We might be tempted to simply take this run and use the combination of factors at those specific levels. However, if we were to look at the full factorial experiment for these seven factors, there would have been 2187 total runs. We have taken a small sample from the larger population of runs, and, therefore, the single run that we obtained that produced the desired value of 100 may not be the best of the 2187! We must perform an analysis to determine the direction to the best of the 2187 runs. This is the principle of fractional factorial methodology.

The data are analyzed as explained in Chapter 7 to obtain both the average response and the signal-to-noise ratio (S/N) for each level of each

Table 9-2.

col:#	METOL 1	H Q 2	Sulf 5	KBr 9	Alk 10	Temp 12	Speed 13	Image Quality
	5.0	13.0	40	1.0	40	70	5.0	0
	5.0	13.0	60	3.0	60	80	6.5	23.5
	5.0	13.0	80	5.0	80	90	8.0	0
	5.0	14.5	40	3.0	60	90	8.0	0
	5.0	14.5	60	5.0	80	70	5.0	0
	5.0	14.5	80	1.0	40	80	6.5	31.0
	5.0	16.0	40	5.0	80	80	6.5	38.5
	5.0	16.0	60	1.0	40	90	8.0	0
	5.0	16.0	80	3.0	60	70	5.0	0
	6.5	13.0	40	3.0	80	80	8.0	38.5
	6.5	13.0	60	5.0	40	90	5.0	0
	6.5	13.0	80	1.0	60	70	6.5	0
	6.5	14.5	40	5.0	40	70	6.5	0
	6.5	14.5	60	1.0	60	80	8.0	50.5
	6.5	14.5	80	3.0	80	90	5.0	0
	6.5	16.0	40	1.0	60	90	5.0	9
	6.5	16.0	60	3.0	80	70	6.5	4.5
	6.5	16.0	80	5.0	40	80	8.0	62.5
	8.0	13.0	40	5.0	60	90	6.5	0
	8.0	13.0	60	1.0	80	70	8.0	0
	8.0	13.0	80	3.0	40	80	5.0	67.0
	8.0	14.5	40	1.0	80	80	5.0	83.5
	8.0	14.5	60	3.0	40	90	6.5	16.5
	8.0	14.5	80	5.0	60	70	8.0	12.0
	8.0	16.0	40	3.0	40	70	8.0	28.5
	8.0	16.0	60	5.0	60	80	5.0	100.0
	8.0	16.0	80	1.0	80	90	6.5	33.0

factor. Note that since there are many responses that measure zero, a constant of 0.01 was added to each response to allow the S/N (B) to be computed. The results of this analysis are found in Table 9-3.

We will now take the information from Table 9-3 and plot it to see the trends that exist between the seven factors and the average IQ and the S/N type B. The type B S/N is used, since we have observed that so many of the test points in the experiment produced very low values and we are trying to

Table 9-3.

	FACTOR	Averages	S/N (B)	
	Metol			
Levels	5.0	10.34	-38.24	LINEAR SUM OF SQ= 3403.125
	6.5	18.34	-36.48	QUADRATIC SUM SQ= 198.375
	8.0	37.84	-33.47	
	H Q			
Levels	13.0	14.34	-38.24	LINEAR SUM OF SQ= 1200.5
	14.5	21.51	-36.48	QUADRATIC SUM SQ= 6.0
	16.0	30.68	-33.47	
	Sulfite			
Levels	40.0	22.01	-36.48	LINEAR SUM OF SQ= 3.125
	60.0	21.68	-36.48	QUADRATIC SUM SQ= 3.375
	80.0	22.84	-36.48	
	KBr			
Levels	1.0	23.01	-36.48	LINEAR SUM OF SQ= 2.0
	3.0	19.84	-35.22	QUADRATIC SUM SQ= 73.5
	5.0	23.68	-36.21	
	Alkali			
Levels	40.0	22.84	-36.48	LINEAR SUM OF SQ= 3.1
	60.0	21.68	-36.48	QUADRATIC SUM SQ= 3.4
	80.0	22.01	-36.48	
	Temperature			
Levels	70.0	5.01	-38.24	LINEAR SUM OF SQ= 10.1
	80.0	55.01	-32.29	QUADRATIC SUM SQ= 14553.4
	90.0	6.51	-38.24	
	Speed			
Levels	5.0	28.84	-37.45	LINEAR SUM OF SQ= 253.1
	6.5	16.34	-35.22	QUADRATIC SUM SQ= 459.4
	8.0	21.34	-36.48	

increase the IQ level. We could have also used the type T S/N and it would have brought us to the same conclusions we are about to reach.

From our interpretation of the graphic representation of the data, we will select the levels of each of the seven factors that maximize both the IQ and the S/N. If there is little or no effect due to a factor, we will select the level for that factor based on other considerations, such as cost or throughput.

The two developer ingredients (hydroquinone and metol) maximize our responses at their high levels. We may be inclined to go to even higher concentrations with these materials, but there is a problem with the solubility of these powders in water much beyond the top levels we have chosen.

The sulfite, KBr, and alkali show very little influence on the IQ and the S/N. We will therefore use the lower end of the concentration range and save money in the formulation of this developer by using less material. Also, as photochemists, we realize that the Br ion is a byproduct of the development process. KBr is included in the developer to prevent spontaneous development of unexposed grains of film. However, as the inhibitor halogen ion (Br) builds up, the development process begins to slow down. To remedy this situation, we will have to periodically dump developer and water it down to keep the Br concentration at the proper level. This process is part of the replenishment operation, and if replenishment can be minimized we will have more up time and greater productivity.

Our experiment has not told us about either of the ideas just discussed, but it did tell us that these factors do not play a part in influencing the IQ response and therefore may be set at levels based on other prior information. This is a very important aspect of the interpretation stage of an experimental design. We will utilize every piece of prior information at every stage of our experimentation. Table 9-4 summarizes the selection of levels we have made based on the plots of Figure 9-1.

We will now try these combinations of factors and their levels in a process trial that produces an IQ of 100, which is right on the value that we desired. Notice that the above combination of values produced the value of 100 and that the trial from the original L27 design also produced a value of 100. If we had used the settings of that particular run from the experiment, we would have set the sulfite and alkali at a noneconomical value of 60 g/l and the KBr would have been both uneconomical and too high from a replenishment point of view at 5 g/l. The analysis of trends is the most important part of the experimental analysis step.

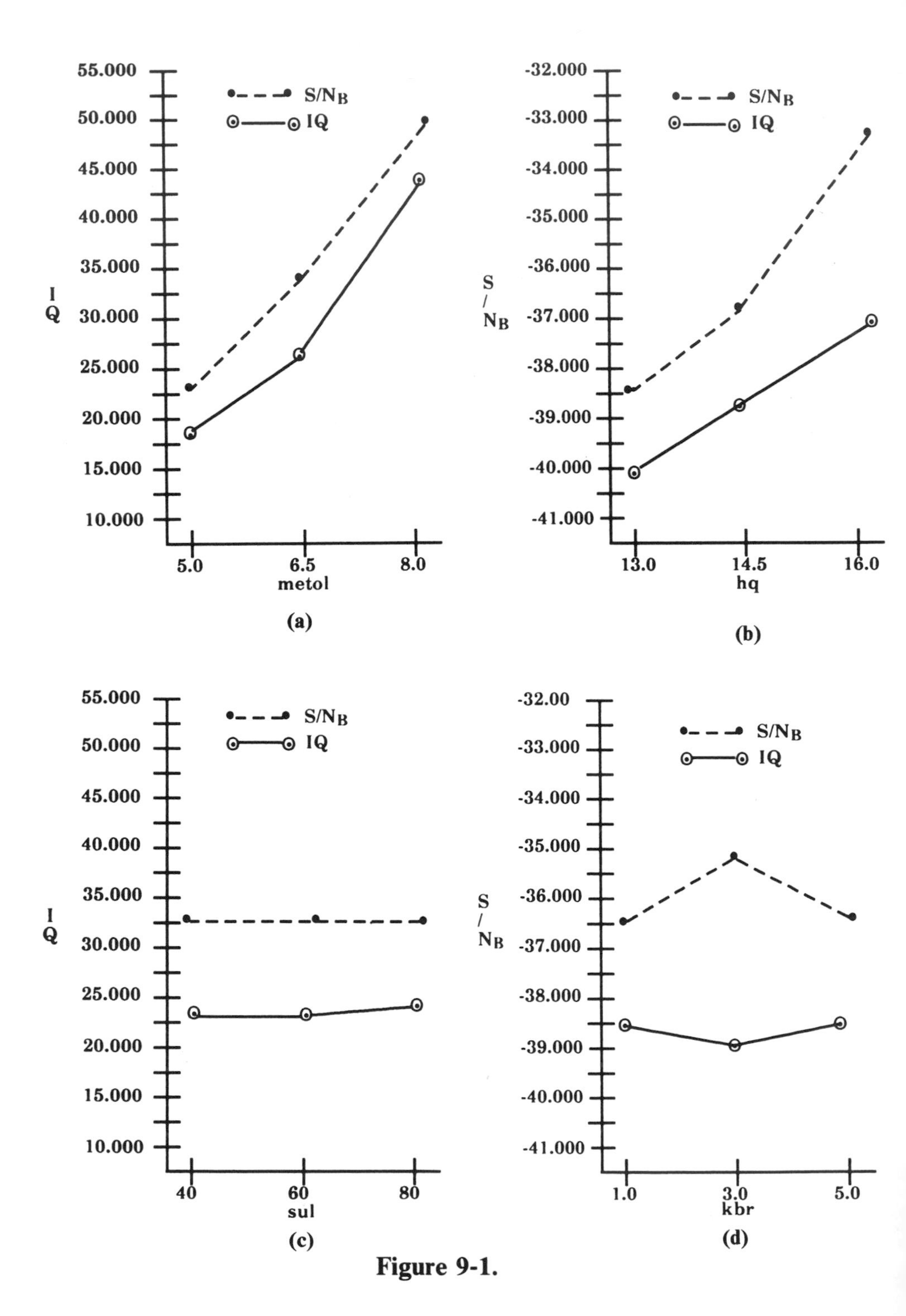

55.000
50.000
45.000
40.000
35.000
30.000
25.000
20.000
15.000
10.000
IQ
S/NB
IQ
5.0
6.5
8.0
metol
(a)
-32.000
-33.000
-34.000
-35.000
-36.000
-37.000
-38.000
-39.000
-40.000
-41.000
S/NB
S/NB
IQ
13.0
14.5
16.0
hq
(b)
55.000
50.000
45.000
40.000
35.000
30.000
25.000
20.000
15.000
10.000
IQ
S/NB
IQ
40
60
80
sul
(c)
-32.00
-33.000
-34.000
-35.000
-36.000
-37.000
-38.000
-39.000
-40.000
-41.000
S/NB
S/NB
IQ
1.0
3.0
5.0
kbr
(d)

Figure 9-1.

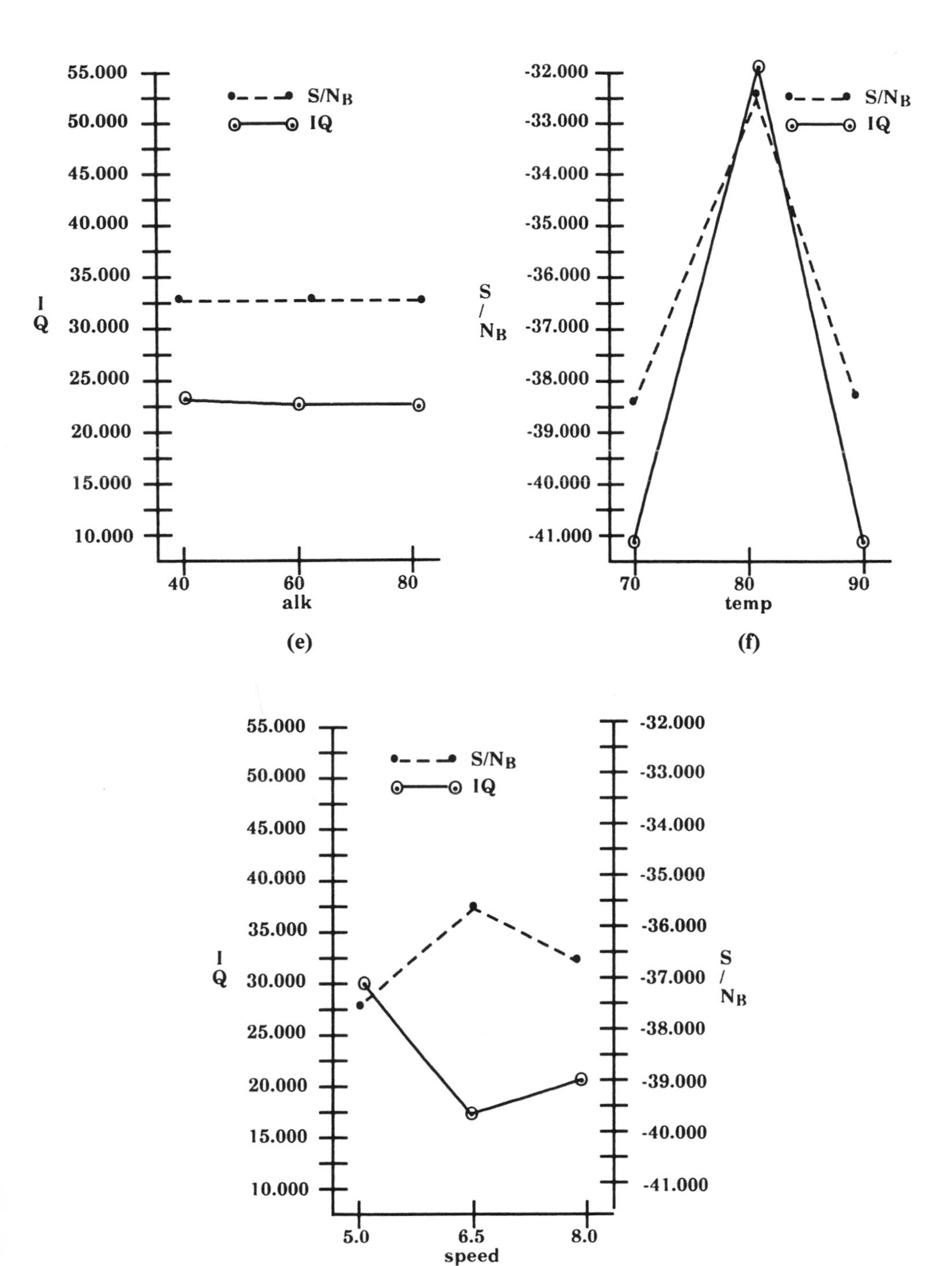

55.000
50.000
45.000
40.000
35.000
30.000
25.000
20.000
15.000
10.000
I Q
S/N_B
IQ
40
60
80
alk
(e)
-32.000
-33.000
-34.000
-35.000
-36.000
-37.000
-38.000
-39.000
-40.000
-41.000
S / N_B
70
80
90
temp
(f)
5.0
6.5
8.0
speed
(g)

Table 9-4.

FACTOR NAME	SELECTED LEVEL
Metol	8 g/l
H Q	16 g/l
Temperature	80 F
Speed	5 ips
KBr	1 g/l
Sulfite	40 g/l
Alkali	40 g/l

The verification *test* (a one-shot trial) showed that we achieved the target of 100. Now we need to see if the variation around the set points will lead to a tolerable variation around the output characteristic, the IQ. To test this, we could run a process capability trial, allowing the seven factors to randomly change, as they naturally would in a typical process. However, as we saw in Chapter 6, a more efficient method to determine capability is to purposely vary the components in a designed experiment. This allows us to make sure that all the variation takes place over a limited number of runs, and also, because of its design structure, we are able to analyze the results and to determine which components are the quality-sensitive components.

We run the structured experimental design, as shown in Table 9-5. We have utilized an L18 experimental structure for this tolerance design. The L18 is used to reduce the amount of effort in this stage of experimentation and is justified, since the manifestation of interactions is minimized as the levels for the factors are brought closer together. We have also used the 1.2247 factor to allow our uniform distribution experiment to produce output statistics that appear to come from a normal distribution.

It is apparent by observing the information in the percent contribution (% CONT.) column of Table 9-6 that there are only two quality-sensitive components in this process. The metol developer accounts for nearly 75% of the variation and the hydroquinone is the remaining 25%. Before we decide on new tolerances for these factors, we should determine if it is even necessary to reduce the variation. We will utilize the loss func-

Table 9-5.

Speed	Temp	HQ	Metol	Sulfite	KBr	Alkali	Image Quality
4.88	79.4	15.69	7.69	38.78	0.76	38.78	91.03
4.88	80	16	8.0	40	1	40	100.36
4.88	80.6	16.31	8.31	41.22	1.24	41.22	109.63
5.0	79.4	15.69	8.0	40	1.24	41.22	96.36
5.0	80	16	8.31	41.22	0.76	38.78	105.90
5.0	80.6	16.31	7.69	38.78	1	40	97.10
5.12	79.4	16	7.69	41.22	1	41.22	93.60
5.12	80	16.31	8.0	38.78	1.24	38.78	103.00
5.12	80.6	15.69	8.31	40	0.76	40	101.70
4.88	79.4	16.31	8.31	40	1	38.78	109.60
4.88	80	15.69	7.69	41.22	1.24	40	91.40
4.88	80.6	16	8.0	38.78	0.76	41.22	100.10
5.0	79.4	16	8.31	38.78	1.24	40	105.66
5.0	80	16.31	7.69	40	0.76	41.22	97.40
5.0	80.6	15.69	8.0	41.22	1	38.78	96.36
5.12	79.4	16.31	8.0	41.22	0.76	40	102.83
5.12	80	15.69	8.31	38.78	1	41.22	101.93
5.12	80.6	16	7.69	40	1.24	38.78	93.60

tion to help make this decision. Our market research shows the following trends in the marketplace.

The information plotted in Figure 9-2 shows the growth trends in our customers' desire for higher quality images over the years. For the purpose of determining how much quality is enough, we will take the current 1990 information and construct a single-sided quadratic loss function. Figure 9-3 shows how the average loss of $12 is linked with the LD50 (50% point) of the return curve. The k for this function is:

$$L(y) = k\,(y - m)^2. \tag{9-1}$$

$$\$12 = k\,(96 - 100)^2. \tag{9-2}$$

$$k = \frac{\$12}{16}. \tag{9-3}$$

$$k = \$0.75. \tag{9-4}$$

Table 9-6. Tolerance Analysis of Developer Formulation.

SOURCE		SUM OF SQ.	DF	% CONT.	AVERAGE EFFECTS LEVEL-1	LEVEL-2	LEVEL-3
Speed	-(L)	2.49341	1	0.45	100.3400	99.7833	99.4283
	-(Q)	0.04067	1	0.01			
Temp	-(L)	0.02803	1	0.01	99.8400	99.9683	99.7433
	-(Q)	0.12484	1	0.02			
H Q	-(L)	139.67360	1	25.00	96.4383	99.8517	103.2617
	-(Q)	0.00001	1	0.00			
Metol	-(L)	416.18740	1	74.50	93.9617	99.8500	105.7400
	-(Q)	0.00000	1	0.00			
Sulfite	-(L)	0.02803	1	0.01	99.8183	99.8183	99.9150
	-(Q)	0.00934	1	0.00			
KBr	-(L)	0.02613	1	0.00	99.8200	99.8183	99.9133
	-(Q)	0.00934	1	0.00			
Alkali	-(L)	0.02708	1	0.00	99.9150	99.8167	99.8200
	-(Q)	0.01034	1	0.00			
RESIDUAL		0.00006	3	0.00			
TOTAL		558.65830	17	100.00	AVERAGE=	99.85 STD. DEV.=	5.73

We now have the required information to compute an expected loss. We must be careful in applying the expected loss formula in this case, since there is no loss above the IQ of 100. To exercise this care, we will compute the entire expected loss and then, given that our process is centered at 100, we can discard the loss on the upper side of the target value. Because of the symmetry of the distribution of IQ (assumed to be normal), we can accomplish this goal, by simply dividing the entire expected loss by two, as shown in the expression in Figure 9-3. Table 9-7 shows the results of these calculations for various IQ standard deviations (SDs).

Our finance and marketing management has targeted no more than 60 cents loss per roll for this problem this year. This financial goal translates to a SD of no greater than 1.25 based on the calculated expected losses in Table 9-7. Armed with this information, we are able to engineer our process to meet the quality goals of the company. As explained in Chapter 6, we now establish the transmission of variation relationship that links the quality-sensitive input components with the final allowable output variation:

$$\frac{1.25^2}{5.73^2} = [(\text{MET})^2 \cdot .745 + (\text{HQ})^2 \cdot .25 + .005], \tag{9.5}$$

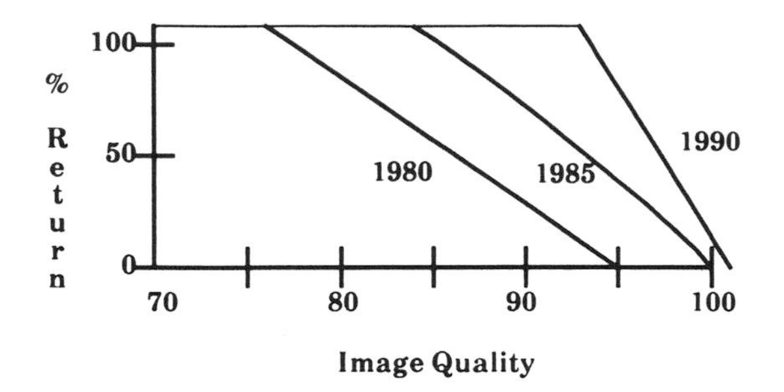

Year	LD 50 for Image Quality	Average Loss per Return
1980	85	\$ 7.00
1985	93	\$10.00
1990	96	\$12.00

Figure 9-2.

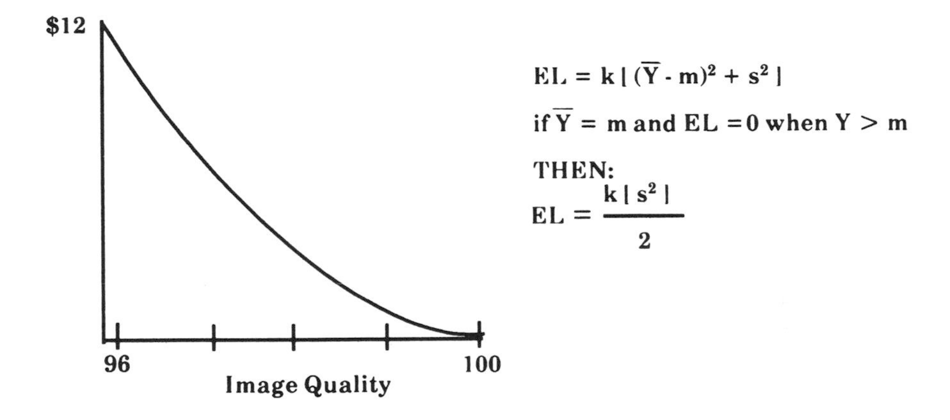

Figure 9-3.

where MET is the fractional reduction of metol and HQ is the fractional reduction of hydroquinone.

We now need to find the values of MET and HQ that will satisfy the equation (i.e., make the right side of the equation equal the left side of the equation). There are many combinations of values that will satisfy the requirements. However, since we will be using the same device to weigh both components, it would be foolish to have different tolerances for the in-

Table 9-7.

Image Quality Standard Deviation	Expected Loss
5.73	$12.31 (result of process study)
4.00	6.00
3.00	3.38
2.00	1.50
1.75	1.15
1.50	0.84
>>>1.25	0.59<<<
1.00	0.38

dividual components. From this consideration, we may combine the metol and hydroquinone fractional reduction factors into a single weighing tolerance factor (WT) and solve for its value:

$$\frac{1.25^2}{5.73^2} = [(\mathrm{WT})^2 \cdot .995 + .005]. \tag{9-6}$$

$$.04659 - .005 = \mathrm{WT}^2 \cdot .995. \tag{9-7}$$

$$.04159 = \mathrm{WT}^2 \cdot .995. \tag{9-8}$$

$$\sqrt{\frac{.04159}{.995}} = \mathrm{WT}. \tag{9-9}$$

$$\mathrm{WT} = .2044. \tag{9-10}$$

Remember, WT is the fractional reduction that we will apply to the tolerable variation in the weighing of the two developer components. The present, nontolerable weighing variation is a SD of 0.25 g/l. To reduce the variation in the final IQ, we will need to reduce the variation of these quality-sensitive components *to* .2044 or about one-fifth of the original variation. Numerically this is a SD of 0.05 g/l.

New variation = old variation · fractional reduction.

For our example,

$$s_{WN} = s_{WO} \cdot WT, \tag{9-11}$$

where s_{WN} is the new weighing SD, s_{WO} is the old weighing SD, and WT is the fractional reduction factor for the weighing operation.

$$s_{WN} = 0.25 \cdot 0.2044. \tag{9-12}$$

$$s_{WN} = 0.0511. \tag{9-13}$$

We are just about finished with our engineering activities in this process. There is one final process capability study that must be made to confirm our predictions that this developer formulation will provide the superior quality images our customers demand. Figure 9-4 shows the histogram of this process capability study based on 50 batches.

The average IQ is very close to the required value of 100. The variation (expressed as a SD) is slightly higher than predicted. This is not an unexpected result, since we have only a sample of the population of IQ values and the sample is subject to variation itself.

Since our process comes very close to the stated financial requirements (the expected loss is about $.75), we will now issue the toleranced manufacturing specifications. Table 9-8 does this in the proper format and completes the design engineering aspects of this activity. Statistical

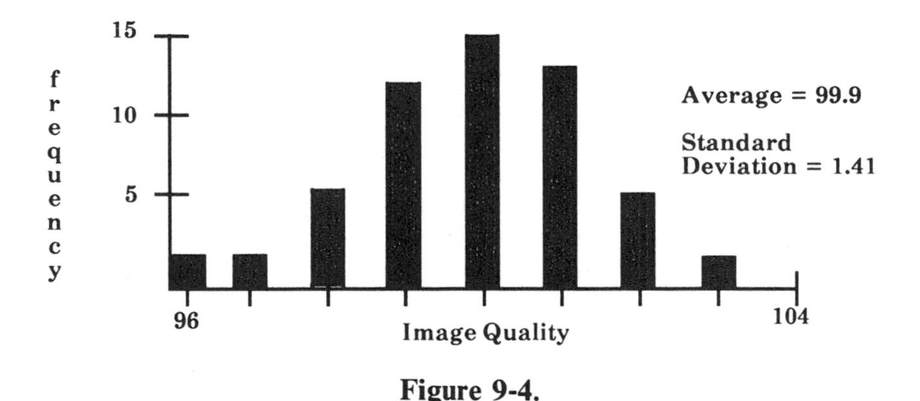

Figure 9-4.

Table 9-8. Issued Specifications.

Parameter Name	Set-point Condition	1 std. dev. Tolerance
Metol	8 g/l	.05 g/l
H Q	16 g/l	.05 g/l
Temperature	80 F	.5 F
Speed	5 ips	.1 ips
KBr	1 g/l	.2 g/l
Sulfite	40 g/l	1.0 g/l
Alkali	40 g/l	1.0 g/l

process control efforts will now take over and ensure a continuing effort in our quest for quality.

AN ELECTROSTATIC COPIER CHARGING DEVICE

In our next example we will look at an electromechanical application of engineering quality by design and also see how these methods help with the interface of subsystems in the broader system/subsystem approach to product development.

Through market research, the voice of the customer has strongly indicated that higher, more consistent IQ is a major consideration in purchasing or leasing copier equipment. The level of IQ is directly related to the optical density or blackness of the image. The level of blackness is related to the electrostatic charge impinged on the photoconductive surface. The photoconductor is the "reusable film" that receives the image we are copying. Figure 9-5 shows a diagram of a typical electrostatic copier device.

Although there are many interrelated aspects of this copier that tie together to produce the level of quality the customer wants, our design team "owns" only a small subsystem in the overall design. We are responsible for the positioning of a bracket that will hold a device that puts the charge on the photoconductor. Figure 9-6 shows our subsystem and the neighboring components.

The angle bracket supports a rail, which is attached with a fastener that threads into a drilled and tapped hole in the top of our bracket. A cor-

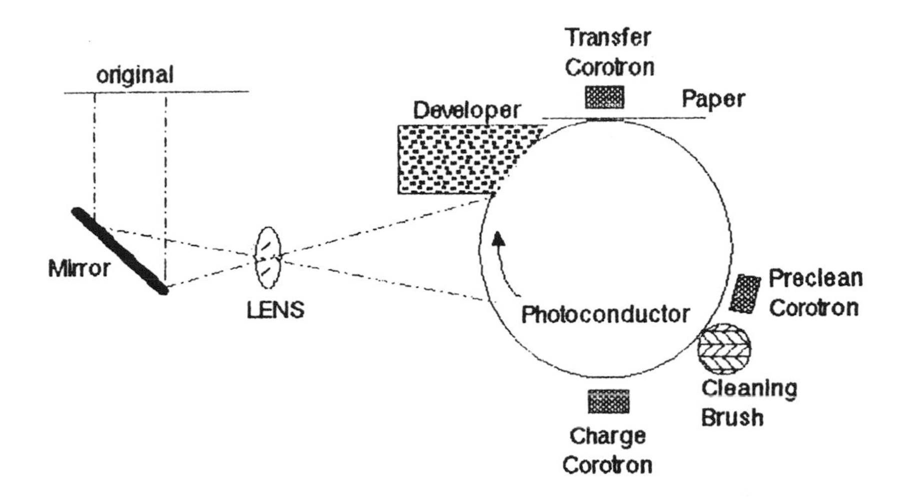

Figure 9-5.

otron (a corona discharge device) box made of aluminum slides on the rail. The corotron wire is mounted in the corotron box and is insulated from ground potential. The wire is connected to a power supply. When the power is turned on, electrons flow from the corotron wire to the surface of the photoconductor, where they will remain as long as the photoconductor is kept in darkness. As soon as light strikes the surface of the photoconductor, the charge dissipates to ground. The lens in Figure 9-5 focuses the light on the photoconductor, and the pattern of light and dark (such as the pattern of words on this page) in the original that is projected by the lens now becomes an electrostatic pattern on the photoconductor. There is a charge remaining where the black letters were in the original and no charge where the white paper was in the original. It is impossible to see the charge visually (it is called a *latent* image), so we next dust the photoconductor with a fine charged powder (called *toner* or *dry ink*). The toner has an opposite charge to that on the photoconductor, so it is attracted to the latent image and we now can see where the charge remains. What we see, of course, is the black pattern from the original.

Since the photoconductor is often rather heavy, inflexible, and expensive, we remove the toner (still in the pattern of the image) from the photoconductor by charging a piece of paper oppositely to the toner and

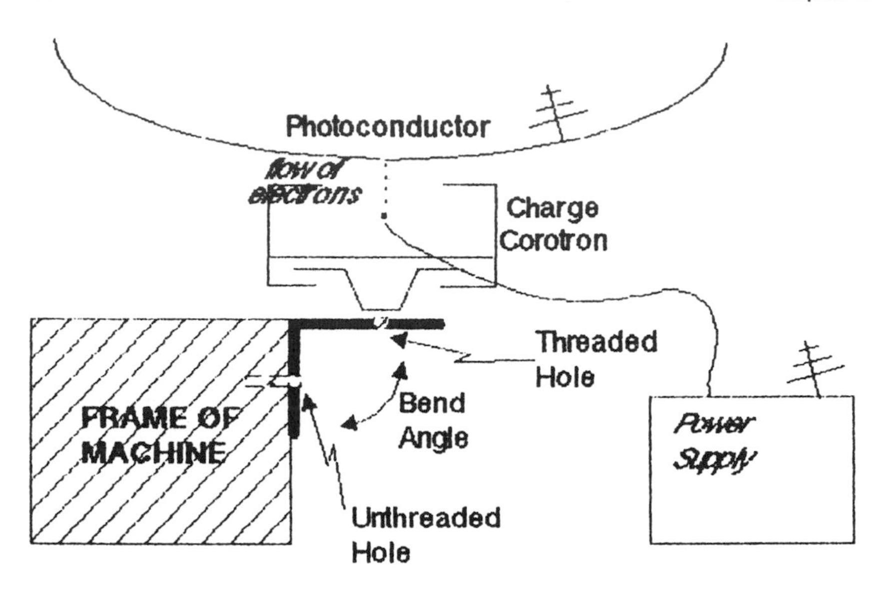

Figure 9-6.

at a higher potential than the photoconductor, and thus transfer the toner to the paper. This charging is done with another corotron (transfer corotron) shown at the top of Figure 9-5 .

There is still a small amount of residual toner on the photoconductor, so its charge is neutralized and then physically cleaned with a rotating brush to make the surface of the photoconductor ready to repeat the cycle all over again. (A *historical note:* The first demonstration of the process described above took place on October 22, 1938 in Astoria, Queens, NY. Chester Carlson, who was the inventor of the process, used a handwritten original "10-22-38 Astoria." It took nearly 20 years to get this basic system design into a product, which was the Xerox 914 machine, and was put on the market in 1959.)

Over the three decades since the Xerox 914 was first introduced, the level of expectation of copy machines has grown and grown. Part of that growth in expectation is related to the project we are about to undertake. We have a goal to produce high-quality and consistent images despite manufacturing variation and despite outside influences that are beyond our design authority.

The engineering team has brainstormed the problem and the following factors shown in Tables 9-9 and 9-10 will be investigated in our experiment.

There are a number of possible approaches to the robustification process. We might consider the inner/outer array method using a three- level design matrix (L27) and a two-level noise matrix (L8). This would lead to 216 individual tests, which are far too many for the short development cycle time period we have.

Another approach often used by Dr. Taguchi is to put only the outside noise factors in the outer arrays and to use only the worst conditions con-

Table 9-9.

Design Authority Factor Name	Working Range	Variation (1 standard deviation)
Bend Angle	88° - 91°	0.5°
Metal Thickness	1.9 - 2.2mm	.01mm
Unthreaded Hole Location	14.5 - 16.mm	.01mm
Threaded Hole Location	14.0 - 16.0mm	.01mm
Torque on Fasteners	90 - 110 inch ounces	2.0 in. oz.

Table 9-10.

Outside Factors		
Power Supply current	20 microamps	.5 microamps
Relative Humidity (RH)	50%	8.0% (RH units)

trasted against the best conditions. Our design team considers midpoint current (20 μA) and 50% RH as ideal conditions. Worst conditions would be low current and high humidity. If we were to pursue this approach, we would run two L27 arrays that included only the five design authority factors. One L27 would be run at ideal conditions and the other L27 would be run at worst conditions. The responses of the two arrays would be averaged and the S/N computed between the two sets of responses. This approach cuts the 216 runs down to only 54 and makes the job possible.

We could also use internal stress method (ISM), as we did in the previous developer problem. This would take only on L27 array. With this method, we would include the external noise factors as part of the experimental configuration.

Table 9-11.

Bend	Metal Thick	Unth Hole	Thrd Hole	Torque	Voltage Best Case	Voltage Worst Case	S/N (T)	Average
88	1.9	14.5	14.0	90	803.5	779	15.29	791.2
88	1.9	15.25	15.0	100	863.9	839.5	22.12	851.7
88	1.9	16.0	16.0	110	931.5	907	28.90	919.2
88	2.05	14.5	15.0	100	825.6	801	17.22	813.3
88	2.05	15.25	16.0	110	889.4	864.9	27.79	877.1
88	2.05	16.0	14.0	90	877.3	852.8	24.70	865.0
88	2.2	14.5	16.0	110	851	826.5	20.16	838.7
88	2.2	15.25	14.0	90	835.2	810.7	18.23	822.9
88	2.2	16.0	15.0	100	899.3	874.8	31.05	887.0
89.5	1.9	14.5	15.0	110	865.3	840.8	22.35	853.0
89.5	1.9	15.25	16.0	90	929.2	904.7	29.66	916.9
89.5	1.9	16.0	14.0	100	917	892.5	33.70	904.7
89.5	2.05	14.5	16.0	90	890.8	866.3	28.22	878.5
89.5	2.05	15.25	14.0	100	874.9	850.4	24.18	862.6
89.5	2.05	16.0	15.0	110	939	914.6	26.69	926.8
89.5	2.2	14.5	14.0	100	836.5	812	18.37	824.2
89.5	2.2	15.25	15.0	110	896.9	872.5	30.24	884.7
89.5	2.2	16.0	16.0	90	964.5	940	21.48	952.2
91	1.9	14.5	16.0	100	930.5	906	29.23	918.2
91	1.9	15.25	14.0	110	914.7	890.2	34.14	902.4
91	1.9	16.0	15.0	90	978.8	954.3	19.46	966.5
91	2.05	14.5	14.0	110	876.3	851.8	24.48	864.0
91	2.05	15.25	15.0	90	936.7	912.2	27.33	924.4
91	2.05	16.0	16.0	100	1004.3	979.8	16.71	992.0
91	2.2	14.5	15.0	90	898.3	873.8	30.70	886.0
91	2.2	15.25	16.0	100	962.2	937.7	21.85	949.9
91	2.2	16.0	14.0	110	950	925.5	24.10	937.7

The team decides to use the worst case-best case approach, and the layout of this design is shown in Table 9-11, along with the voltage responses, average voltages, and S/N values.

We complete an analysis of the above data by finding the average at each level of each factor. These results are found in Table 9-12. Since the L27 is a confounded experimental configuration, it is inappropriate to consider the investigation of interactions. We plot these results in Figures 9-7 through 9-11.

Table 9-12.

FACTOR	Voltage	S/N (T)
Bend Angle		
88	851.8	22.8
89.5	889.3	26.1
91	926.8	25.3
Metal Thickness		
1.9	891.5	26.1
2.05	889.3	24.2
2.2	887.0	24.0
Unthreaded Hole		
14.5	851.9	22.9
15.25	888.1	26.2
16.0	927.9	25.2
Threaded Hole		
14.0	863.9	24.1
15.0	888.1	25.2
16.0	915.9	24.9
Torque		
90	889.3	23.9
100	889.3	23.8
110	889.3	26.5

Our first observation is that metal thickness (Figure 9-8) and torque (Figure 9-11) have very little effect on the voltage. However, the S/N is strongly influenced by these factors. We will select the levels that maximize S/N for these two factors. Metal thickness is selected at 1.9 mm and torque is selected at 110-in. oz.

There is a quadratic effect on the S/N from the other three factors, as well as a strong linear effect on the voltage. We will run a trial using the peak S/N points from these factors. Table 9-13 shows the proposed set points for our bracket design and the resulting voltage. Since this voltage is slightly too high (to obtain the proper optical density, a 900-V electrostatic charge is required) we will drop the voltage by changing one of the factors that influence voltage but does not have an adverse effect on the S/N. The factor that fits this requirement best is the threaded hole location. In Figure 9-10 we can see that the slope of the S/N is far less than the slopes of the S/Ns in Figures 9-7 and 9-9, while there is still a strong influence of the threaded hole on the voltage. The relationship is close enough to a linear trend to allow a linear interpolation to determine the setting of the threaded hole that will produce a voltage that is closer to the specified 900 V.

From Figure 9-10 we interpolate a hole location that will drop the voltage by 1.4. The observed voltage at 14 mm is 863.9 and the observed voltage at 15 mm is 888.1. This amounts to a 24.2 V/mm change. Since we only need a 1.4-V change, the threaded hole needs to be changed by only .05 mm. The next trial is shown in Table 9-14 and comes very close to the specified 900 V required for proper image density.

Now that we have "zeroed in" on the proper levels of our design, we need to see what will happen when the tolerances are applied around these levels and what will be the result of variation in the outside noise factors on the charge voltage. This is the tolerance design stage of our investigation. As in the previous example with the photographic developer, we will utilize a structured approach to determining the capability of the design. The statistical structure will be an L18 configuration, and we will include the 1.2247 factor to allow the output voltage results to appear to have been produced by normal distribution inputs. Table 9-15 shows this design with the 18 resulting voltages.

While the average value of these 18 runs is right where we predicted, the standard deviation is far too high to produce the low-variation optical density we require for our customers. Other research has told us that a SD of 5 Vs or less is necessary to keep the optical density variation from becoming noticeable.

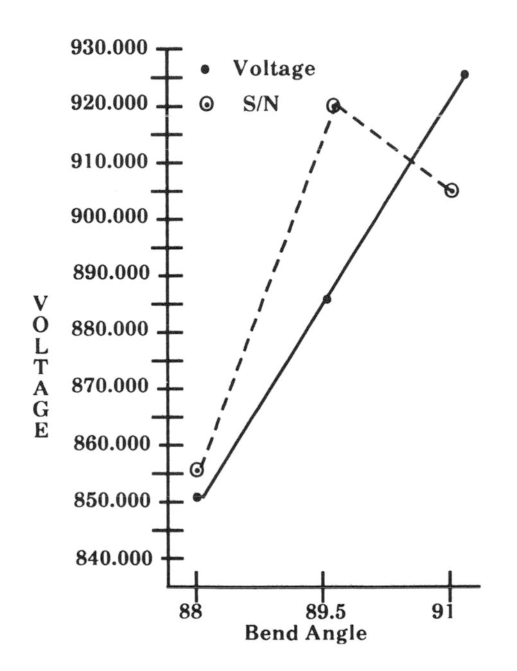

Figure 9-7.

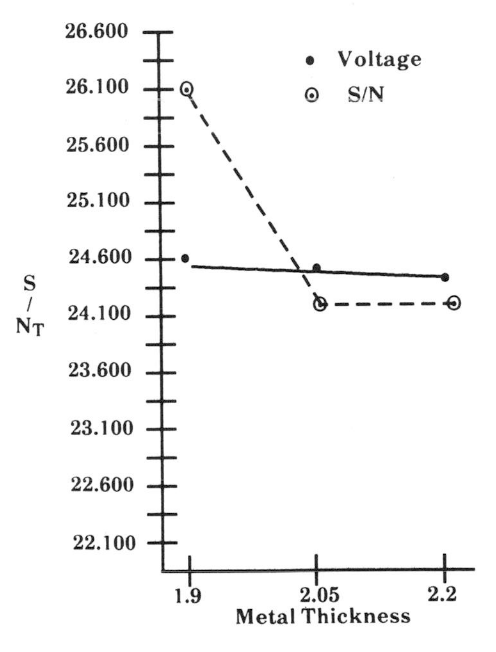

Figure 9-8.

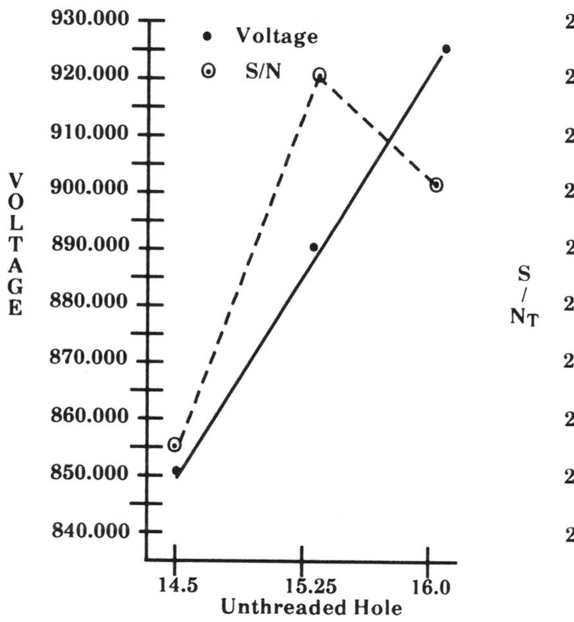

Figure 9-9.

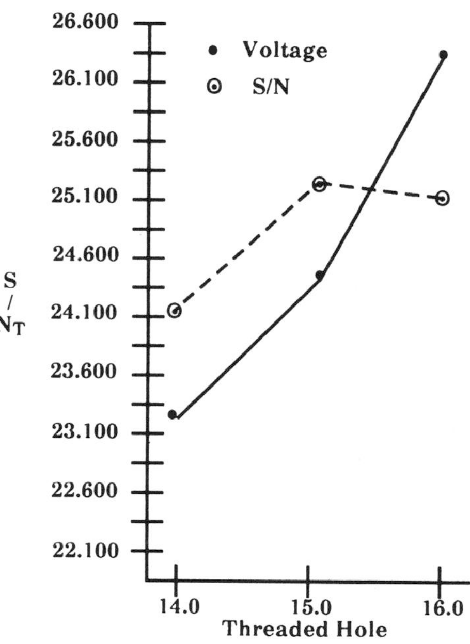

Figure 9-10.

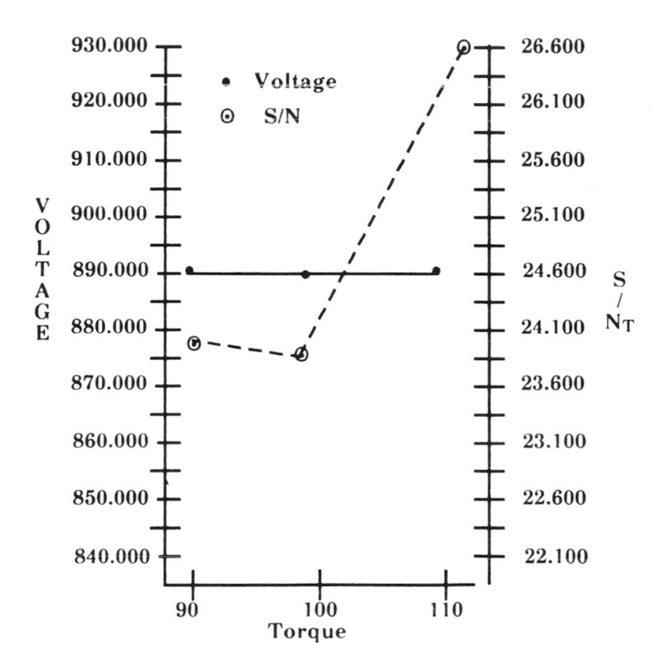
930.000
920.000
910.000
900.000
890.000
880.000
870.000
860.000
850.000
840.000
VOLTAGE
Voltage
S/N
26.600
26.100
25.600
25.100
24.600
24.100
23.600
23.100
22.600
22.100
S/NT
90
100
110
Torque

Figure 9-11.

Table 9-13.

FACTOR NAME	SELECTED LEVEL
Bend Angle	89.5°
Metal Thickness	1.9 mm
Unthreaded Hole	15.25 mm
Threaded Hole	15.00 mm
Torque	110 inch ounces

The run will be at the ideal conditions for current and RH.	20 microamps 50% RH

The resulting voltage was 901.4

Table 9-14.

FACTOR NAME	SELECTED LEVEL
Bend Angle	89.5°
Metal Thickness	1.9 mm
Unthreaded Hole	15.25 mm
Threaded Hole	14.95 mm
Torque	110 inch ounces

The run will be at the ideal conditions for current and RH.	20 microamps 50% RH

The resulting voltage was 900.2

We now run an analysis of the variation sources from the tolerance design in Table 9-15. These results are found in Table 9-16 and indicate that there are two quality-sensitive components, the bend angle and the power supply.

We will now set up the trade-off equation to determine how much to reduce the variations in the bend angle and power supply current. Note that one of these quality-sensitive components (power supply current) is outside of our design authority. We will first determine the extent of its variation reduction and then discuss this very important point. Expressions 9-14 through 9-20 show the details of our rational reduction of variaiton.

$$\frac{5^2}{26.6^2} = [(BA)^2 \cdot .2397 + (PSC)^2 \cdot .7535 + .0068], \qquad (9\text{-}14)$$

Table 9-15.

Bend	Thick	Un TH	Th HL	Torque	Currnt	RH	Volts
88.9	1.89	15.24	14.94	107.6	19.4	40	859
88.9	1.90	15.25	14.95	110.0	20.0	50	885
88.9	1.91	15.26	14.96	112.5	20.6	60	911
89.5	1.89	15.24	14.95	110.0	20.6	60	925
89.5	1.90	15.25	14.96	112.5	19.4	40	875
89.5	1.91	15.26	14.94	107.6	20.0	50	900
90.1	1.89	15.25	14.94	112.5	20.0	60	913
90.1	1.90	15.26	14.95	107.6	20.6	40	946
90.1	1.91	15.24	14.96	110.0	19.4	50	887
88.9	1.89	15.26	14.96	110.0	20.0	40	888
88.9	1.90	15.24	14.94	112.5	20.6	50	911
88.9	1.91	15.25	14.95	107.6	19.4	60	855
89.5	1.89	15.25	14.96	107.6	20.6	50	928
89.5	1.90	15.26	14.94	110.0	19.4	60	870
89.5	1.91	15.24	14.95	112.5	20.0	40	902
90.1	1.89	15.26	14.95	112.5	19.4	50	889
90.1	1.90	15.24	14.96	107.6	20.0	60	913
90.1	1.91	15.25	14.94	110.0	20.6	40	945
						AVERAGE:	900.1 volts
						Std. Dev.:	26.6 volts
						S/N (T):	30.6 dB

Table 9-16. Analysis of Variation Sources from Coroton Tolerance Design.

SOURCE		SUM OF SQ.	DF	% CONT.
Bend	-(L)	2852.08300	1	23.97
	-(Q)	0.02778	1	0.00
Thick	-(L)	0.08333	1	0.00
	-(Q)	0.02778	1	0.00
Unt HL	-(L)	6.75000	1	0.06
	-(Q)	0.02778	1	0.00
Th HL	-(L)	1.33333	1	0.01
	-(Q)	0.11111	1	0.00
Torq	-(L)	0.08333	1	0.00
	-(Q)	0.69444	1	0.01
Currnt	-(L)	8965.33300	1	75.35
	-(Q)	0.11111	1	0.00
R H	-(L)	70.08334	1	0.59
	-(Q)	0.69444	1	0.01
RESIDUAL		0.33398	3	0.00
TOTAL		11897.78000	17	100.00

where BA is the fractional reduction of the bend angle tolerance and PSC is the fractional reduction of the power supply tolerance. We know that it is possible to make a more precise bend if we abandon the break-bend process and opt for the more expensive extrusion/milling approach. This new method will allow the bend angle tolerance a SD of 0.1°. Since our current bend angle tolerance is 0.5°, we are able to reduce this tolerance to one-fifth (0.20) of its present value. We enter this reduction factor into expression 9-14.

$$\frac{5^2}{26.6^2} = [(0.20)^2 \cdot .2397 + (PSC)^2 \cdot .7535 + .0068]. \qquad (9\text{-}15)$$

$$0.03533 = [0.009588 + (PSC)^2 \cdot .7535 + .0068]. \qquad (9\text{-}16)$$

We can now solve for the value of the power supply reduction, since there is only one unknown left in expression 9-16.

$$0.03533 - 0.009588 - 0.0068 = (\text{PSC})^2 \cdot 0.7535. \quad (9\text{-}17)$$

$$0.018942 = (\text{PSC})^2 \cdot 0.7535. \quad (9\text{-}18)$$

$$\frac{0.018942}{0.7535} = (\text{PSC})^2. \quad (9\text{-}19)$$

$$\text{PSC} = .15855. \quad (9\text{-}20)$$

We now multiply the current tolerance value of the power supply by the PSC reduction factor to obtain the new required tolerance of this device. The new SD is reduced from 0.5 μA to 0.08 μA. The question that now arises is if such a tight tolerance is possible. We will have to talk with the design group that is responsible for the power supply and convey our requirements to them. We are their customer, just as the corotron design team is our customer. What we have, in fact, is a series of interfaces between customers and vendors. Figure 9-12 illustrates this general interface concept and Figure 9-13 brings this concept into our particular corotron bracket example.

Each customer/vendor entity represents a subsystem in the larger final system that is the product. If we attempted to construct the whole system and to experiment on it in its entirety, we would never be able to complete the effort, for it would be too vast. Therefore, we have the subsystems. The link between the subsystems is the balance between customer requirements and vendor noise. What is an output of one subsystem is a noise to the next subsystem.

Using Taguchi's concept of outer noise, we are able to link the interfaces and to infuse a continuity throughout the entire system. The concept that a vendor produces noise that a customer must either robustify against or, as in our example, respecify, is a major concept that Taguchi has contributed to the science and engineering of product design. This concept is an answer to the complaint heard so many times, "our company is unique since our processes are so complex that we can't fit them into experimental designs. . . "

Let's look at how this important concept works in our corotron bracket example. Our team is the vendor/customer in the middle of Figure 9-13. We deliver charge voltage to the copy machine, which is our customer.

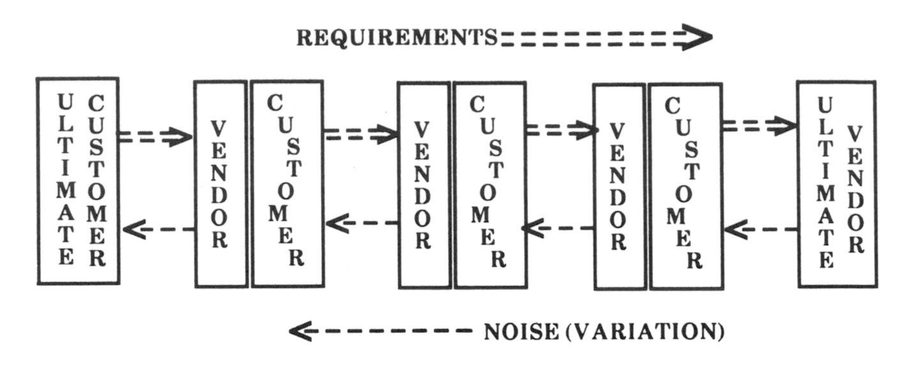

Figure 9-12.

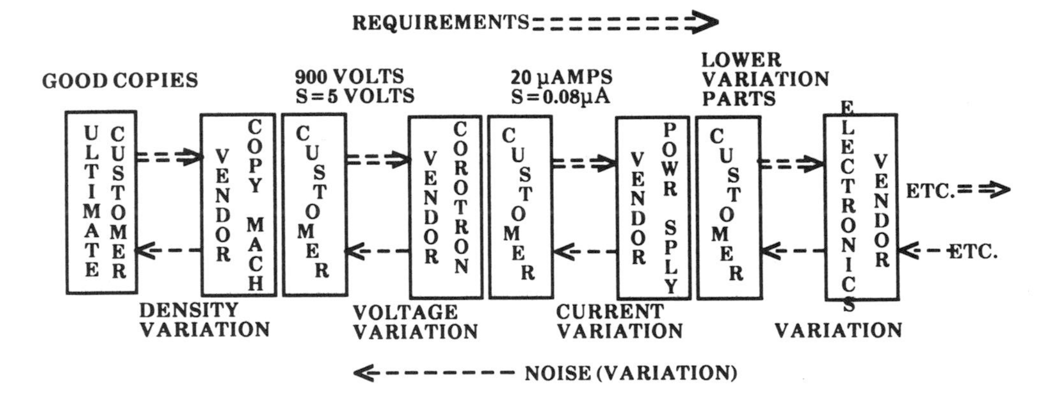

Figure 9-13.

The copy machine delivers copies to the ultimate customer. The ultimate customer drives the entire chain by demanding "good copy quality." The copy machine team determines that good copy quality translates into a charge voltage of 900 V with a SD of 5 V. By the method our corotron bracket team positions the charging device, we are able to produce levels of output voltage. Our output is noise to the next team in the chain. To keep the noise down, we utilize the methods of experimental design and especially parameter design.

However, even with the best parameter design we are sending too much noise to our customer (the copy machine team). The copy machine team has done its best within the physical constraints of its system to withstand the noise we vend to them, for they have also utilized parameter design! If we send them too much noise, they will send their customer too much noise.

Therefore, just as the copy machine team came to us with their customer requirement, we go to our vendor (the power supply team) and state our requirements. The power supply team follows the same path as we and our customer have. The power supply team utilizes the parameter design to robustify their basic system and tolerance design to determine the quality-sensitive components. The sequence of customer/vendor interfaces continues with the customer driving the vendor and the vendor using the methods of engineering quality by design.

We will now run the process capability study for our corotron bracket susbsystem based on the results of the verification/tolerance analysis. Table 9-17 shows the set points as well as the allowable manufacturing and use variation stated in SD units.

The conditions in Table 9-17 confirm our prediction and, more importantly, produce a device that will be able to meet the customer needs. The engineering quality by design process has put us on target, while allowing us to minimize variation and to do so with minimal cost increases, since we have tighter and more costly tolerances only on the components that need such attention.

PLASTIC MOLDING PROCESS

We will look at one more example that uses the more "traditional" approach to engineering quality by design. This example has fewer factors, and we will be working over a more contracted range of these factors. For this reason we will be able to utilize a two-level experimental structure,

Table 9-17.

FACTOR NAME	SELECTED LEVEL	ALLOWABLE VARIATION (Standard Deviation)
Bend Angle	89.5 °	0.1 °
Metal Thickness	1.9 mm	0.01 mm
Unthreaded Hole	15.25 mm	0.01 mm
Threaded Hole	14.95 mm	0.01 mm
Torque	110 inch ounces	2 inch ounces
Current	20 microamps	0.08 microamps
Humidity	50% RH	8% (RH units)

The resulting voltage was 900.2 volts
The standard deviation is 4.95 volts
The type T S/N 45.2 dB

rather than the broader ranged three-level structures used in the previous two examples. Because there are fewer factors and we are using only two levels, we will can employ the inner/outer array method for this problem. The choice of this approach is, of course, because it is appropriate to the situation, as determined by the design team.

Our current molding process is producing plastic parts according to the settings shown in Table 9-18. The parts are adequate, but not of the quality we would like to produce.

The response variable that determines the quality of the molded part is called the molding index (MIdx). It quantifies the degree of incompleteness of the part (a negative index) or the amount of flash or extra plastic on the part (positive index). Figure 9-14 shows this concept diagrammatically.

We have established a loss function for this process based on the cost of replacing incorrectly made parts. When the MIdx drops to 3 or rises to 3 (±3 from perfect), the loss is $.90. Using the quadratic loss function for this situation, we obtain a k of $.10. The expected loss for the current process, which is tending toward short shots and has a large SD, is $.16. This means that we must include in the price of every part made under the current process, $.16 for warranty purposes. The competition is presently

Table 9-18.

FACTOR	SET-POINT	Std. Dev.
Temperature	525	25
Pressure	1100	50
Time	0.85	.01
Gate Size	0.25	.03
Resin Melt Index	17.5	2.5

Molding Index (MIdx)= -0.875
Std. Dev. of MIdx = 0.92

underpricing us by $.14 per part. If we can reduce the warranty costs, we will be more price competitive.

The project team meets in both the brainstorming and rationalization meetings and determines that the factors we presently have specified are the only factors that need to be investigated in the parameter design. They have formulated a goal as follows:

GOAL: To find the system settings that will prevent short shots or flash despite molding machine variation and raw material fluctuations.

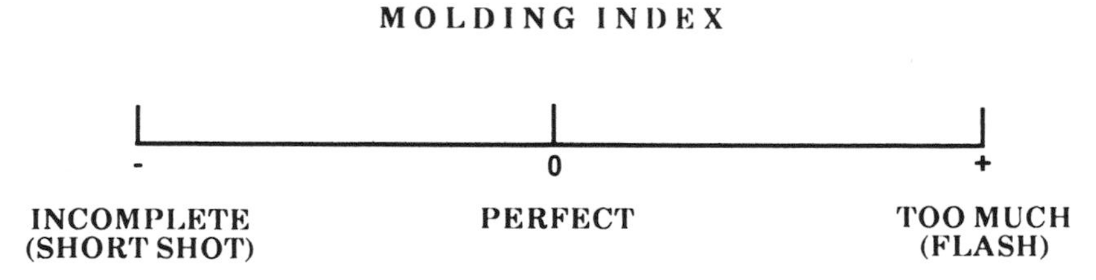

Figure 9-14.

Table 9-19 shows the factors, working ranges, and tolerance values to be used in the experiment. Since there are only four factors and since we expect interactions to exist between these factors, we will use an L16 design configuration for the design matrix (inner array). Since this is a full factorial configuration, columns 8, 4, 2, and 1 are selected. Table 9-20 shows this design.

The team decides that, since the processing conditions can be set quite precisely during an experiment, and since the molding machine is easily reconfigured, they will use the noise matrix (outer array) approach to generating the variation in the parameter design. They pick the L8 as the smallest design capable of including five factors. Columns 4, 2, 1, 6, and 5 are assigned to the noise matrix configurations. The first noise matrix is shown in Table 9-21.

While in this first noise matrix, all of the parts come up as short shots, we still continue on with the experiment. We are not trying to make parts for sale, we are making good information so that we may make better parts later. The remaining noise arrays are run over the next few days and the results are compiled in Table 9-22.

As with our other examples, we will find the average MIdx and average S/N (T) at the low and high levels for each factor and plot these results below.

We will pick the levels of the factors that have the highest S/N values. However, the time factor shows very little influence on the S/N. We can use it as an adjustment (or signal factor) to bring the level of MIdx on to

Table 9-19.

FACTOR	WORKING RANGE	Std. Dev.
	o	o
Temperature	500-550	25
Pressure	1000-1200 psi	50 psi
Time	0.85-1.2 sec.	.01 sec
Gate Size	0.20-0.3 in.	.03 in.
Resin Melt Index	17.5	2.5

Table 9-20.

tc	TEMP	PRESS	TIME	GATE
(1)	500	1000	.85	.2
a	550	1000	.85	.2
b	500	1200	.85	.2
ab	550	1200	.85	.2
c	500	1000	1.2	.2
ac	550	1000	1.2	.2
bc	500	1200	1.2	.2
abc	550	1200	1.2	.2
d	500	1000	.85	.3
ad	550	1000	.85	.3
bd	500	1200	.85	.3
abd	550	1200	.85	.3
cd	500	1000	1.2	.3
acd	550	1000	1.2	.3
bcd	500	1200	1.2	.3
abcd	550	1200	1.2	.3

Table 9-21.

Run#	TEMP	PRESS	TIME	GATE	MELT INDEX	MIdx
1	475	950	.84	.23	20	-2.7
2	525	950	.84	.17	15	-2.4
3	475	1050	.84	.17	20	-1.7
4	525	1050	.84	.23	15	-1.4
5	475	950	.86	.23	15	-4.1
6	525	950	.86	.17	20	-1.8
7	475	1050	.86	.17	15	-3.1
8	525	1050	.86	.23	20	-0.8

MIdx Average = -2.25
S/N (T) = 11.6

Table 9-22.

tc	TEMP	PRESS	TIME	GATE	MIdx	S/N(T)
(1)	500	1000	.85	.2	-2.25	11.6
a	550	1000	.85	.2	-0.75	18.4
b	500	1200	.85	.2	-0.25	18.6
ab	550	1200	.85	.2	1.25	15.5
c	500	1000	1.2	.2	-0.50	18.6
ac	550	1000	1.2	.2	1.00	17.1
bc	500	1200	1.2	.2	1.50	14.1
abc	550	1200	1.2	.2	3.00	9.5
d	500	1000	.85	.3	-2.25	11.8
ad	550	1000	.85	.3	-0.75	19.3
bd	500	1200	.85	.3	-1.75	13.4
abd	550	1200	.85	.3	-0.25	20.8
cd	500	1000	1.2	.3	-0.50	19.6
acd	550	1000	1.2	.3	1.00	17.8
bcd	500	1200	1.2	.3	0.00	19.8
abcd	550	1200	1.2	.3	1.50	14.8

Table 9-23.

FACTOR	MIdx	S/N (T)	FACTOR	MIdx	S/N (T)
Temperature			Time		
500	-.75	16.0	.85	-.87	16.2
550	.75	16.7	1.20	.87	16.4
Pressure			Gate		
1000	-.62	16.8	0.20	.38	15.4
1200	.63	15.8	0.30	-.38	17.2

the target of zero. We may use our data to determine the setting of time for this purpose. Here is the reasoning:

1. The 550 setting of temperature increases the MIdx by 0.75 (+.75).
2. The 1000 setting of pressure decreases the MIdx by .62 (+.13).
3. The .3 setting for gate decreases the MIdx by .38 (−.25).
4. To overcome this MIdx drop of .25, we need to hold the time at a point that will add .25 to the MIdx. A linear interpolation on the time factor gives us a setting of 1.075 to achieve this requirement.

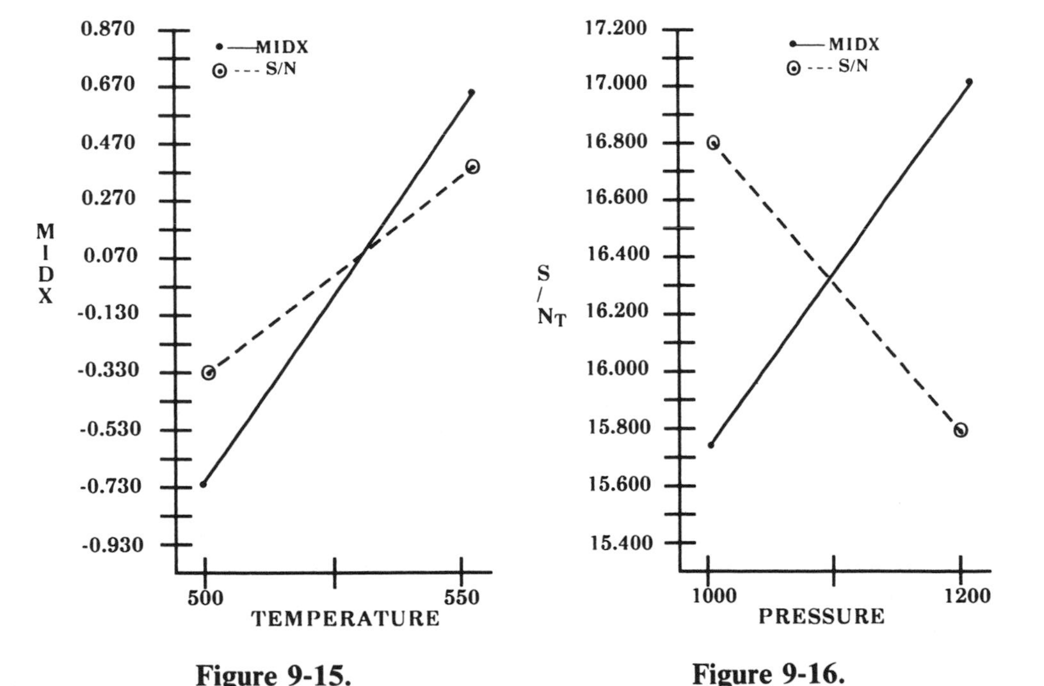

Figure 9-15.

Figure 9-16.

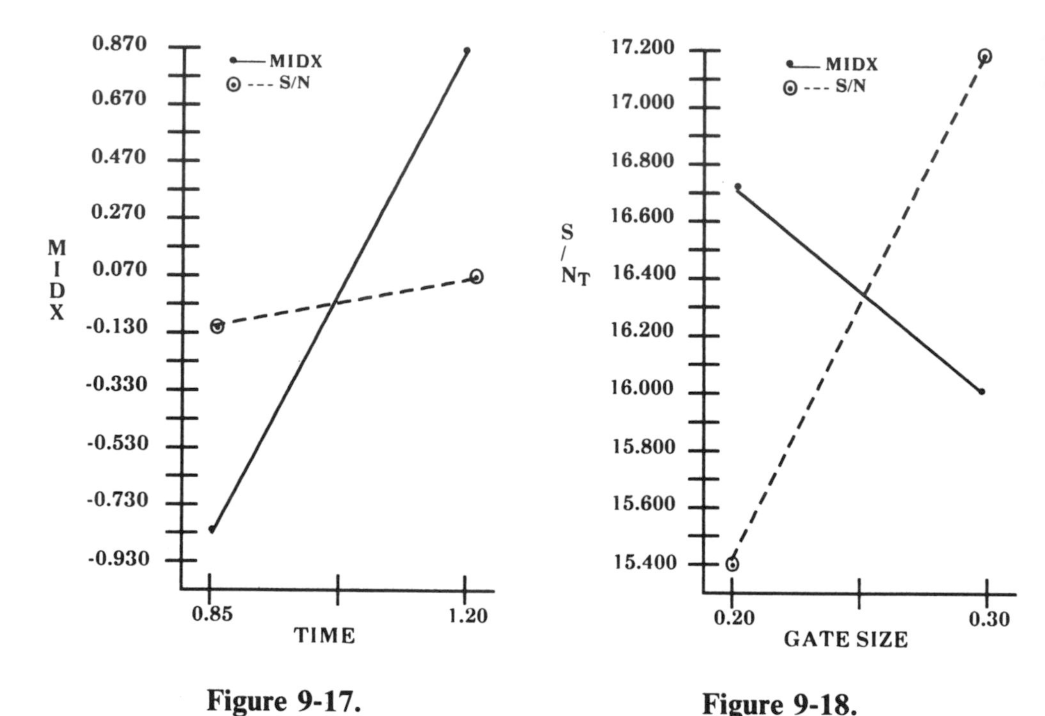

Figure 9-17.

Figure 9-18.

When we try this set of conditions out, we are surprised that the result is a positive .38 MIdx! Was our reasoning incorrect? The reasoning is correct, but it is not complete. There is another effect that we have ignored in the graphical interpretation of the data. We have not looked at the possibility of interactions between the factors. In the presence of interactions, the single-effects plots do not tell the entire story. A simple statistical analysis tells us that there is a significant interaction between the gate and the pressure. This interaction is plotted in Figure 9-19 and shows both why our predicted MIdx was off and why the S/N led us to the levels of pressure and gate size.

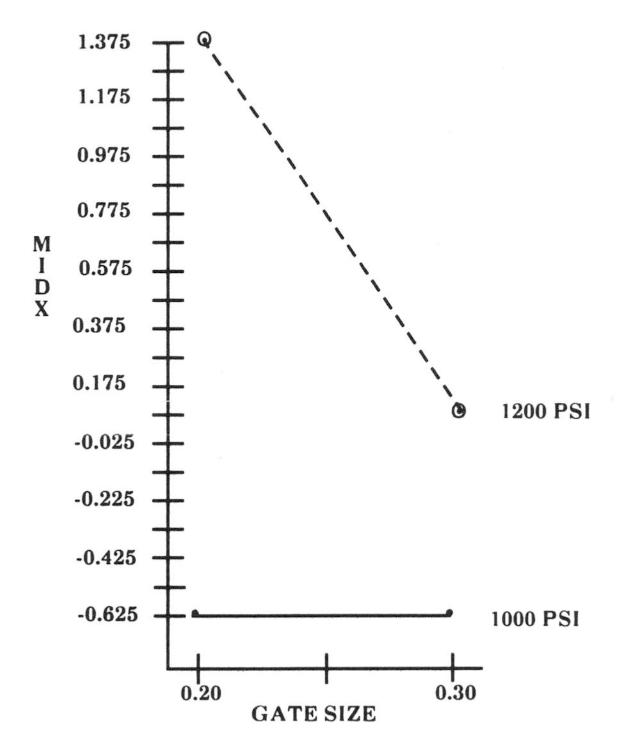

Figure 9-19.

Let us first comment on why the S/N selected the levels of gate and pressure. Figure 9-19 shows that if we use the low (1000 psi) pressure, there is no effect on the MIdx as the gate size changes. Also, the difference in MIdx as a function of pressure with a gate of .3 in. is far smaller than with a gate of .2 in. The robust levels are the levels that transmit less variation to the response. The robust levels produce lower variation and therefore higher S/N values. The interaction was the driver in the S/Ns selection of the gate and pressure levels.

Why was our prediction of the MIdx off the mark? In our prediction we looked at gate and pressure separately. They do not influence MIdx separately. If we look at their combined effect at the correct setting in the interaction graph, we see that the overall drop in MIdx is only .62, not the sum of the drop due to the individual, average influence we first reported based on the single-effect plots. We thought there was an overall drop in MIdx of 1.0, but it was only .62. The difference is the +.38 we observed in our verification test run. It is important to always look at the interaction before looking at single effects. However, it is reassuring to note that if we run a verification test of the conditions and we do not see the result we expected, it is possible to reinterpolate until we obtain the value we require. Of course, it is much faster and less costly to utilize all the data before running many test runs that hone the settings for us. It is better to use thinking than just brute force!

We now have the correct settings that will put us on target, but we need to see how the variation around these settings influences the variation in the MIdx output. We will utilize the tolerance design methods at this point. As before, we set up a tolerance design, and the results for the original tolerances are shown in Table 9-24.

The variation we have encountered is somewhat better than we experienced in our initial production runs (it was 0.92). However, the major contributor to the loss has been eliminated, since we are now on the correct target value and we know why we are there! Now a simple rational reduction of variation is in order based on the contribution analysis shown in Table 9-25.

We will need to control the temperature and the gate size to meet our loss requirements. If we reduce the temperature variation to 20% of its

Table 9-24.

FACTOR	SET POINT	Std. Dev.
	o	o
Temperature	550	25
Pressure	1000 psi	50 psi
Time	1.0 sec.	.01 sec
Gate Size	0.3 in.	.03 in.
Resin Melt Index	17.5	2.5

Average MIdx = 0
Std. Dev. = 0.83

Table 9-25. Tolerance Analysis.

FACTOR	% CONTRIBUTION		
Temperature	85.8%	(QSC)	
Pressure	2.4%		Required output
Time	0.4%		MIdx variation
Gate	4.8%	(QSC)	is 0.30
Melt	---		to meet loss
Unexplained	6.6%		requirements.

current value and the gate variation to 33% of its current value, we will achieve this goal.

Table 9-26 shows the final process design and its output. We are still on the target of zero MIdx, and the observed SD is 0.22. This is a lower variation than we had predicted. Looking back to the tolerance analysis, we see that there was a residual of 6.6%. This was due to the interaction we had already seen in the parameter design. When we reduced the variation in the gate, we also took out some of the variation due to the interaction with the gate, and in doing so, we cut the overall variation of the system. This is a synergistic effect that sometimes accompanies such efforts and another reason for understanding interactions.

Table 9-26.

FACTOR	SET POINT	Std. Dev.
	o	o
Temperature	550	5
Pressure	1000 psi	50 psi
Time	1.0 sec.	.01 sec
Gate Size	0.3 in.	.01 in.
Resin Melt Index	17.5	2.5

Average MIdx = 0
Std. Dev. = 0.22

While it would be nice to have examples for all possible situations where engineering quality by design is to be applied, it would be impossible. Each situation is unique in its requirements. Therefore, each situation must be approached on its own characteristics and the proper method applied. Using the wrong tool of investigation can ruin the effort. The examples in this chapter have been provided not so much as "templates," but more as guides to the thinking that goes into designing experiments.

One of the most exciting parts of the discipline of experimental design is the variety. An experimental designer never gets bored due to this variety. There is an irony in this thought, since we design experiments to eliminate output variation. Variation is the nemesis of quality, but variety is the spice of life!

PROBLEMS FOR CHAPTER 9

1. Verify the calculations of the loss function k value for the plastic molding example and draw the quadratic curve. Then compute the expected loss for the initial process (data in Table 9-18) and for the final process (data in Table 9-25). For this plastics process, will we be able to match the competition in price? Could we beat their price?
2. An automobile engine has a specified horsepower of 107 with a SD of 2. Propose an approach to accomplishing this goal using engineering quality by design methods. The factors identified by the design team are

Factor	Working range	SD of set point
Compression ratio	7–12	.3
Valve timing (BTDC)	30–45°	1°
Spark advance (BTDC)	10–20°	.5°
Fuel flow (lbs/hr)	.06–18	20%
rpm	600–6000	5%
Atmospheric moisture	20–100 grains	20 grains
Octane of fuel	87–92	.05

3. Pick a product design or process from your line of work and set up a table of factors and levels, and tolerances that is similar to the table in question #2 above. Consider how you would attack this problem using engineering quality by design.

FURTHER READING

1. Molding the "Impossible Part." *Plastics Technol.* January, 1989, 74.
2. Pao, T. W., Phadke, M.S., and Sherrerd, C. S. Computer response time optimization using orthogoanal array experiments. IEEE International Communications Conference. Chicago, IL (June 23-26, 1985) Conference Record, vol. 2, 890-895.
3. Bendell, A., Disney, J., Pridmore, W. A., Eds. *Taguchi Methods: Applications in World Industry.* Springer-Verlag, New York, 1989.

10
Orthogonal Array Structures for Experimental Designs

He who distrusts Taguchi indiscriminately will be ignorant unnecessarily.

He who accepts Taguchi indiscriminately will be misled unnecessarily.

There is a great deal of controversy surrounding the Taguchi approach to engineering quality. The subject is debated on an international scale (1). Most of the debate finds its origin among the scientific and mathematical professionals who argue that Taguchi has overly simplified experimental design techniques to the point that the simplification becomes harmful. They further propose that Taguchi's analysis is "overly complicated" and could be accomplished using more conventional approaches (2). The other side of the controversy (those who have successfully used the methods of Taguchi) have a tendency to ignore the scientific arguments and to get on with their business. It is unfortunate that both sides lose in such a debate.

The community of the scientists "know what they know" and try to interpret Taguchi's new ideas in the strictness of that knowledge. The practitioners know little about the roots of the methods and simply follow the rules to produce successful results. There is a blind distrust on the part of the theorists and a blind following on the part of the users. As was shown in Chapter 4, the experimental design structures do have a fundamental origin in the formal methods of experimental design. Taguchi did not invent experimental design structures, but has packaged this information in easy-to-use tabular form.

DOES HISTORY REPEAT ITSELF?

Before we continue and show Taguchi's orthogonal Arrays (OAs), it is interesting to note that during the 19th century an electrical engineer, Heaviside, led and encouraged scientists and engineers to solve complex differential equations by using "operator methods." The advantage was a simplification that allowed the solution of the differential to take place within the normal rules of algebra. It was remarkable to the more rigorous mathematicians that this method led to correct answers. However, the debate that erupted from mathematicians who did not like to see such "blind" applications rewarded with success produced some of the same rhetoric that we hear associated with Taguchi's simplification of experimental design configurations. "Do you need to understand the process of digestion to eat?," was a remark then, as it is now.

Of course the Laplace transforms placed Heaviside's manipulations on an honorable, rigorous foundation. We have not yet found the Laplace for Taguchi, but we do indeed find the Taguchi approach to experimental design easier to use and capable of producing correct answers. This chapter will present a selection from his tabulated OAs and present a strategy for the use of these design structures and their associated linear graphs to best fulfill their promise of success.

THREE-LEVEL STRUCTURES

Since three-level designs are valuable contributors to quality optimization and are also tedious to construct from the classical approach, we will look at the three OAs that are practical from a resource allocation viewpoint. The three designs are the L9, L18, and L27. While there are other OAs (L36, L54, L81, and more), these are simply just too large for ordinary industrial applications.

Recall from Chapter 4 that OAs are balanced, fractional factorial designs. The protocol Taguchi has utilized to name the various OAs describes the capabilities of these designs. We will review that now.

We will look at the L9 design, since it is the smallest and easiest to relate back to its origins.

```
L            99999
L           9     9
L           9     9
L             9 9 9
L                 9
L                9
L               9
LLLLLLL        9
```

The *L* stands for *Latin square,* which is the origin of the discovery of fractional factorial designs.

The *9* stands for the number of runs in the entire experiment.

In general, the name of the experimental structure follows this format:

Character position	Character	Meaning
1st	L	Latin-square based
2nd	#	Number of runs in the design
3rd	#	Number of levels
4th	Exponent #	Maximum number of factors

Therefore the L9 is a Latin-square-based, 9-run, three-level, four-factor structure. The use of a design like this with its maximum number of factors will lead to incorrect results in certain situations, and a strategy for proper use is necessary. We will present such strategies for each design we include in this chapter, but will go into more detail with the L9 to show the basis for these strategies.

As we saw in Chapter 4, confounding (or the superimposition of one effect upon another) takes place in fractional factorial designs. It is this confounding that actually allows us to fractionate the larger design to the size that we can afford to run. When the confounding gets to the point that likely effects are inseparable, we have a case of a poor experimental structure. To prevent using poor experimental structures, we need a guide to the

Table 10-1. L9 (3^4) Strategy Table.

Run Number \ Column number:	1	2	3	4
1	1	1	1	1
2	1	2	2	2
3	1	3	3	3
4	2	1	2	3
5	2	2	3	1
6	2	3	1	2
7	3	1	3	2
8	3	2	1	3
9	3	3	2	1

LINEAR GRAPH FOR L9

STRATEGY for use: Place factor A in column 1
Place factor B in column 2

IF NO INTERACTIONS ARE POSSIBLE BETWEEN ALL FACTORS

THEN: Place factor C in column 3
Place factor D in column 4

confounding patterns. In Chapter 4 we saw how the linear graphs provide this guide to confounding.

In the L9, the linear graph shows that columns 1 and 2 are at the end of the line and that columns 3 and 4 are on the line. Recall that this means that if one places a factor in column 1 and another factor in column 2, the interaction (if it exists between these two factors) will be analyzed by looking to columns 3 and 4. To gain a physical understanding of the last statement, let us look at the actual 3 × 3 matrix layout of the L9 and observe how we would treat the data.

Table 10-2 shows a typical L9 experiment with the data from the nine runs. When we determine the effect of factor A, we find the average change in the response for each level of factor A. At the bottom of Table 10-2 we show the three averages (5, 7, 9). We perform a similar averaging for factor B, but in its case we average across the columns to obtain the row averages found on the right in Table 10-2. Of course, we may perform these averaging operations because the design is balanced (each level of each factor is equally represented across each level of the other factor).

Table 10-2.

		FACTOR A			
		1	2	3	AVERAGE FOR B:
FACTOR B	1	4	5	6	5
	2	5	7	9	7
	3	6	9	12	9
AVERAGE FOR A:		5	7	9	

However, we may also wish to see if there are any nonadditive effects with our responses. To display the nonadditivity, we plot the change in the response as a function of one of the factors, but instead of averaging over the other factor, as we did in Table 10-2 to obtain averages at the right and bottom margins of the table, we will isolate each level of the other factor. This interaction plot technique is simply the *family of curves* method that is well known to scientists and engineers.

When we plot the data from Table 10-2 as a family of curves, we see that the slopes of the lines are not the same. If the slopes differ, we describe this condition as an interaction. The interaction requires the use of the nine individual points in the L9 array. These nine points have positions in the array described by the combination of the levels of the factors. These combinations of levels comprise the subscripts of the matrix and are found in columns 3 and 4 of the L9. This is what we mean by the statement "the interaction will be analyzed by looking to columns 3 and 4."

Now look at Table 10-3. We have brought the 3 × 3 matrix together with the orthogonal array L9. The row, column subscripts of the 3 × 3 ma-

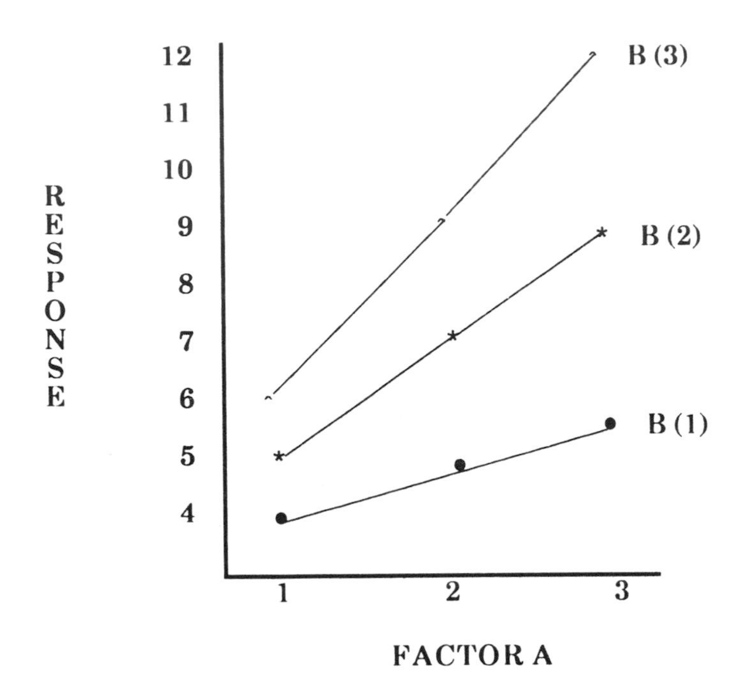

Figure 10-1.

Table 10-3.

FACTOR B	FACTOR A 1	2	3
1	1,1 !	1,2 ^	1,3 *
2	2,1 ?	2,2 @	2,3 $
3	3,1 %	3,2 &	3,3 #

L9 ORTHOGONAL ARRAY

Column number: 1	2	3	4	
1	1	1	1	!
1	2	2	2	@
1	3	3	3	#
2	1	2	3	$
2	2	3	1	%
2	3	1	2	^
3	1	3	2	&
3	2	1	3	*
3	3	2	1	?

trix have been identified by nine symbols that have been matched with the column pairs found in the L9 orthogonal array. This matching shows how the information needed to understand the interaction between the factors in columns 1 and 2 comes from the information in columns 3 and 4.

So, if there is an interaction between the factors in columns 1 and 2, the information in columns 3 and 4 depends on what is physically happening in a nonadditive (interactive) way due to the factors in columns 1 and 2. If we place other factors in columns 3 and 4, we observe the physical effects of those other factors, as well as the physical effect of the interaction influence of the factors in columns 1 and 2. We become confused (or confounded) with regard to the actual effects of the factors in columns 3 and 4, because we observe the total linear combination of the effect of the single factor as well as the interaction.

For example, if the effect of the single factor placed in column 3 is −5 and the effect of the interaction is +5, we observe the sum of these effects, which is *zero!* We say nothing is happening, when two things are happening!

The strategy for using the L9 as shown in its strategy table is to use only columns 1 and 2 if we suspect interactions between any of the factors under investigation. We may only use all four columns of the L9 if *all* interactions may be ruled out.

The same thought process has been carried out for the remaining *strategy tables* in this chapter. The concepts hold for both two- and three-level designs, and are based on the underlying mathematics of group theory.

What if it is necessary to run an L9 experiment with more than two factors and we know that there will be an interaction? There is a further strategy that applies to all the OAs. The important point is that we really know about the existence of the interaction. Table 10-3 is an illustration of such a situation. Let us suppose that we are interested in determining the optical density (blackness) of a photographic film. We want to look at three factors in this experiment and need three levels to be able to look at possible curves. Two of the factors are known to interact. This prior knowledge is shown in Table 10-4 and plotted in Figure 10-2. The interaction is known in the photographic science and engineering discipline as *reciprocity law failure* and is the non-additivity of the exposure intensity and exposure time.

If we use the L9 for this problem and place the intensity in column 1 and the time in column 2, we will have their interaction superimposed on the temperature factor that would go into column 3. To allow this experi-

Table 10-4.

Exposure Time	Exposure Intensity .5	1.0	2.0
2	.20	.28	.48
4	.25	.44	.80
8	.40	.75	1.40

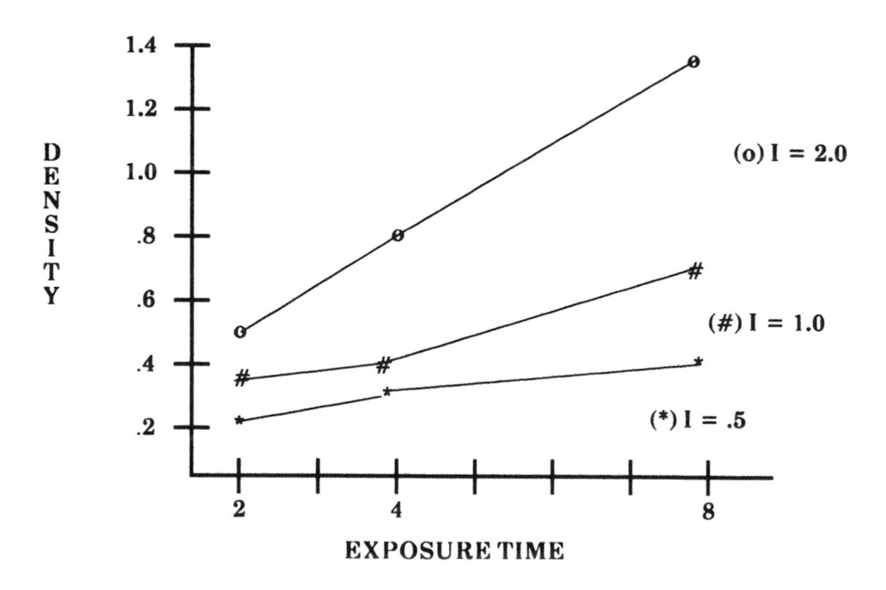

Figure 10-2.

ment to operate within the framework of nine runs, we will utilize our prior knowledge of the time-intensity interaction and *combine* these factors into a more fundamental factor, which in this case is *energy*. Table 10-5 shows this concept and Table 10-6 shows the two-factor L9 arrangement for this particular experiment. The combining factors approach is an appropriate method to utilize prior knowledge and to prevent confounding from biasing our experiments. We do lose the information on the interaction in this experiment, but we already knew the interaction from other research. The more we know, the less we need to do!

L27

The L27 design is a frequently used three-level design. It has the same origins as the L9, but is far more complex due to the number of runs and the number of possible effects. There are many linear graphs for the L27, however, the triangular linear graph shown with the strategy for its use is the one linear graph that allows the most factors to be included in this experimental structure. It is possible to generate customized linear graphs from the following interaction Table 10-7.

Table 10-5.

Time \ Intensity	.5	1.0	2.0
2	ENERGY LEVEL 1 (EL1)		
4		ENERGY LEVEL 2 (EL2)	
8			ENERGY LEVEL 3 (EL3)

Table 10-6.

	Exposure Energy	Developer Temperature
column:	1	2
run		
1	EL1	70
2	EL1	75
3	EL1	80
4	EL2	70
5	EL2	75
6	EL2	80
7	EL3	70
8	EL3	75
9	EL3	80

To use this table, find the intersection of the columns that contain factors that are suspected to interact. At that intersection, the two columns that are confounded with the interaction are found. For example, if we have placed factors in columns 1 and 2, the linear graph has told us that the interaction will appear in columns 3 and 4. We obtain this same information by finding the intersection of column 1 and column 2 in the above interaction table.

Using this table is much like using a table on a road map to find the distance between cities. Another example of its use would be for an interaction between columns that are not on the linear graph. Let's say that we place two factors in columns 9 and 10, but these factors influence the response in a nonadditive manner via an interaction. The interaction table (Table 10-7) shows that the intersection of column 9 with column 10 places this interaction in columns 1 and 8. We are not concerned with column 8, since we have not placed a factor there if we have followed the strategy, but column 1 is one of our prime columns! A part of the column 9-10 interaction has been confounded with the factor in column 1. This unwanted result shows why the strategy is to keep the likely interactions in columns 1, 2, and 5, and out of columns 9, 10, 12, and 13.

The last of the three-level designs that we will present is the L18. It is a hybrid design that is not capable of being derived using the ordinary algorithm that produced the L9 or the L27. It is essentially two L9s with three extra three-level columns plus a two-level column. It can be a very alluring design, because of this mixed-level feature.

L27 (3^{13}) Strategy Table.

Column number:	1	2	3	4	5	6	7	8	9	10	11	12	13
Run Number	o	o	x	x	o	x	x	x	o	o	x	o	o
1	1	1	1	1	1	1	1	1	1	1	1	1	1
2	1	1	1	1	2	2	2	2	2	2	2	2	2
3	1	1	1	1	3	3	3	3	3	3	3	3	3
4	1	2	2	2	1	1	1	2	2	2	3	3	3
5	1	2	2	2	2	2	2	3	3	3	1	1	1
6	1	2	2	2	3	3	3	1	1	1	2	2	2
7	1	3	3	3	1	1	1	3	3	3	2	2	2
8	1	3	3	3	2	2	2	1	1	1	3	3	3
9	1	3	3	3	3	3	3	2	2	2	1	1	1
10	2	1	2	3	1	2	3	1	2	3	1	2	3
11	2	1	2	3	2	3	1	2	3	1	2	3	1
12	2	1	2	3	3	1	2	3	1	2	3	1	2
13	2	2	3	1	1	2	3	2	3	1	3	1	2
14	2	2	3	1	2	3	1	3	1	2	1	2	3
15	2	2	3	1	3	1	2	1	2	3	2	3	1
16	2	3	1	2	1	2	3	3	1	2	2	3	1
17	2	3	1	2	2	3	1	1	2	3	3	1	2
18	2	3	1	2	3	1	2	2	3	1	1	2	3
19	3	1	3	2	1	3	2	1	3	2	1	3	2
20	3	1	3	2	2	1	3	2	1	3	2	1	3
21	3	1	3	2	3	2	1	3	2	1	3	2	1
22	3	2	1	3	1	3	2	2	1	3	3	2	1
23	3	2	1	3	2	1	3	3	2	1	1	3	2
24	3	2	1	3	3	2	1	1	3	2	2	1	3
25	3	3	2	1	1	3	2	3	2	1	2	1	3
26	3	3	2	1	2	1	3	1	3	2	3	2	1
27	3	3	2	1	3	2	1	2	1	3	1	3	2

Linear Graph for L27.

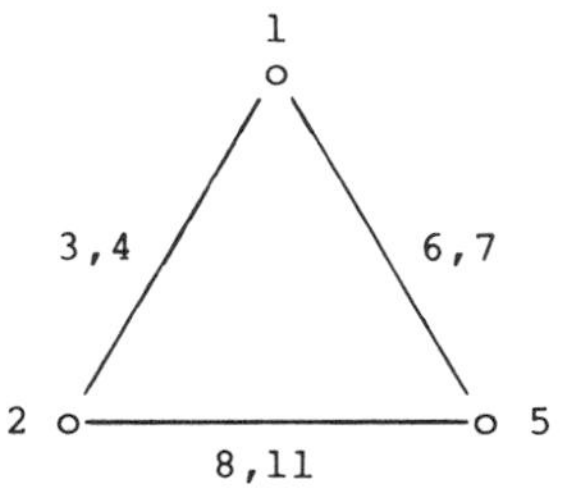

NOTE: o Means OK to use column.
x Means do not use, involved with interactions from other columns.

9 10 12 13

o o o o

STRATEGY for use:							
Place Factor:	A	B	C	D	E	F	G
In Column:	1	2	5	9	10	12	13

Cautions:	Do not waste these columns.	Factors D through G may not interact with any factors.
Strategy:	Place the factors that are most likely to interact with each other in columns 1,2,5.	Place the factors that are least likely to interact with any factors in columns 9,10,12,13.

Table 10-7. Interaction Table for L27.

	2	3	4	5	6	7	8	9	10	11	12	13
1	3,4	2,3	2,3	6,7	5,7	5,6	9,10	8,10	8,9	12,13	11,13	11,12
	2	1,4	1,3	8,11	9,12	10,13	5,11	6,12	7,13	5,8	6,9	7,10
		3	1,2	9,13	10,11	8,12	7,12	5,13	6,11	6,10	7,8	5,9
			4	10,12	8,13	9,11	6,13	7,11	5,12	7,9	5,10	6,8
				5	1,7	1,6	2,11	3,13	4,12	2,8	4,10	3,9
					6	1,5	4,13	2,12	3,13	3,10	2,9	4,8
						7	3,12	4,11	2,13	4,9	3,8	2,10
							8	1,10	1,9	2,5	3,7	4,6
								9	1,8	4,7	2,6	3,5
									10	3,6	4,7	2,7
										11	1,13	1,12
											12	1,11

However, beware of the L18, since it has a very involved pattern of confounding.

In the L-18 the interaction effects are dispersed over all of the three-level columns. This means that no column is safe from confounding if there are any interactions among the effects we study in our experiment.

THE L18 MAY ONLY BE USED IF WE KNOW THERE ARE NO INTERACTIONS BETWEEN ANY OF THE FACTORS.

L18 (3^7, 2^1) Strategy Table.

Run Number \ Column number:	1	2	3	4	5	6	7	8
1	1	1	1	1	1	1	1	1
2	1	2	2	2	2	2	2	1
3	1	3	3	3	3	3	3	1
4	2	1	1	2	2	3	3	1
5	2	2	2	3	3	1	1	1
6	2	3	3	1	1	2	2	1
7	3	1	2	1	3	2	3	1
8	3	2	3	2	1	3	1	1
9	3	3	1	3	2	1	2	1
10	1	1	3	3	2	2	1	2
11	1	2	1	1	3	3	2	2
12	1	3	2	2	1	1	3	2
13	2	1	2	3	1	3	2	2
14	2	2	3	1	2	1	3	2
15	2	3	1	2	3	2	1	2
16	3	1	3	2	3	1	2	2
17	3	2	1	3	1	2	3	2
18	3	3	2	1	2	3	1	2

LINEAR GRAPH FOR L18

```
1            8
o------------o
```

STRATEGY for use:	Place Factor:	A	B	C	D	E	F	G	H
	In Column:	1	2	3	4	5	6	7	8

Cautions: None of the three level factors may interact with any of the other three level factors.

Strategy: Use only in situations where the interaction effects are minimal, or nonexistent.

NOTE: While the linear graph correctly shows no possible confounding between the factor in column 1 and the factor in column 8, this condition exists only because if these are the only factors in the experiment, the L18 becomes a full factorial design.

TWO-LEVEL STRUCTURES

In Chapter 4 we showed the development of the two-level experimental design configurations. We will present three of Taguchi's OAs for two-level designs and relate these structures to the more "classical" approach taken in the Western world.

As we move from the very simple L4 on this page to the more complex L16, we will see that the strategies for the use of these designs become more and more complex. There are a greater number of linear graphs for the larger designs, and these linear graphs also become very involved and detailed. The *o* and *x* notation is not a part of Taguchi's presentation, and has been provided in this book to supplement the interpretation of his linear graphs and to help the experimenter with his or her design strategy.

L4 (2^3) Strategy Table.

Column number:	1	2	3		"WESTERN EQUIVALENT"		
Run Number	o	o	x		B	A	AB
1	1	1	1		-	-	+
2	1	2	2		-	+	-
3	2	1	2		+	-	-
4	2	2	1		+	+	+

Defining Contrast: 1=ABC

LINEAR GRAPH FOR L4

```
1     3     2
o-----------o
```

NOTE: o Means OK to use column
x Means do not use, involved with interactions from other columns.

STRATEGY for use:	Place Factor:	A	B	C
	In Column:	1	2	3

Cautions: In the presence of interactions, this design will fail to produce correct information. Good for only 2 factors.

Strategy: Assign factors A and B to columns 1 and 2. Assign factor C only to column 3 if there are no interactions among all the factors.

L8 (2^7) Strategy Table.

"WESTERN EQUIVALENT"

Column number:	1	2	3	4	5	6	7	1	2	3	4	5	6	7
Run Number	o	o	x	o	x	x	o	C	B	-BC	A	-AC	-AB	ABC
1	1	1	1	1	1	1	1	-	-	-	-	-	-	-
2	1	1	1	2	2	2	2	-	-	-	+	+	+	+
3	1	2	2	1	1	2	2	-	+	+	-	-	+	+
4	1	2	2	2	2	1	1	-	+	+	+	+	-	-
5	2	1	2	1	2	1	2	+	-	+	-	+	-	+
6	2	1	2	2	1	2	1	+	-	+	+	-	+	-
7	2	2	1	1	2	2	1	+	+	-	-	+	+	-
8	2	2	1	2	1	1	2	+	+	-	+	-	-	+

LINEAR GRAPH FOR L8

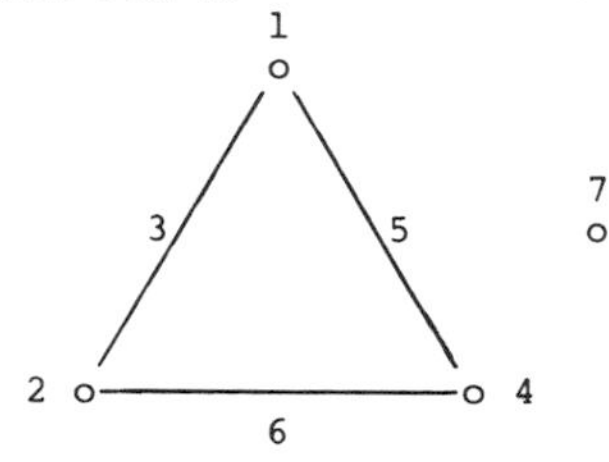

NOTE: o Means OK to use column
x Means do not use, involved with interactions from other columns.

Defining Contrast: 1=ABCD

STRATEGY	Place Factor:	A	B	C	D
for use:	In Column:	1	2	4	7

Cautions: In the presence of interactions, this design will fail to produce correct information if used for more than 4 factors. Good for only 4 factors and the two-factor interactions will be confounded with each other.

Strategy: Assign factors A and B to columns 1 and 2. Assign factor C and D to columns 4 and 7.

STRATEGIES & DEFINING CONTRASTS for other configurations with more factors:
(assign the factor to the column number in the table)

number of factors	A	B	C	D	E	F	G	DEFINING CONTRAST
5	1	2	4	3	5			1=-BCD,-ACE,ABDE
6	1	2	4	3	5	6		1=-BCD,-ACE,-ABF,ABDE, ACDF,BCDF,-DEF
7	1	2	4	3	5	6	7	1=-BCD,-ACE,-ABF,ABCG,ABDE,ACDF,-ADG,BCEF, -BEG,-CFG,-DEF,CDEFG,BDFG,AEFG,-ABCDEFG

NOTE: This L8 design is often wrongly used to study 7 factors in only 8 runs. This may be done only if ALL interactions are negligible. In Western design of experiments terminology, such a design would be called a "saturated design" since all of the information from the experiment goes into the measurement of the single effects. If there are any interactions, they are confounded with ALL of the single effects.

L16 (2^{15}) Strategy Table.

Column number:	1	2	3	4	5	6	7	8	9	10	11	12	13	14	15
Run Number															
1	1	1	1	1	1	1	1	1	1	1	1	1	1	1	1
2	1	1	1	1	1	1	1	2	2	2	2	2	2	2	2
3	1	1	1	2	2	2	2	1	1	1	1	2	2	2	2
4	1	1	1	2	2	2	2	2	2	2	2	1	1	1	1
5	1	2	2	1	1	2	2	1	1	2	2	1	1	2	2
6	1	2	2	1	1	2	2	2	2	1	1	2	2	1	1
7	1	2	2	2	2	1	1	1	1	2	2	2	2	1	1
8	1	2	2	2	2	1	1	2	2	1	1	1	1	2	2
9	2	1	2	1	2	1	2	1	2	1	2	1	2	1	2
10	2	1	2	1	2	1	2	2	1	2	1	2	1	2	1
11	2	1	2	2	1	2	1	1	2	1	2	2	1	2	1
12	2	1	2	2	1	2	1	2	1	2	1	1	2	1	2
13	2	2	1	1	2	2	1	1	2	2	1	1	2	2	1
14	2	2	1	1	2	2	1	2	1	1	2	2	1	1	2
15	2	2	1	2	1	1	2	1	2	2	1	2	1	1	2
16	2	2	1	2	1	1	2	2	1	1	2	1	2	2	1
# factors: 4	o	o	x	o	x	x	x	o	x	x	x	x	x	x	x
# factors: 5	o	o	x	o	x	x	x	o	x	x	x	x	x	x	o
# factors: 6	o	o	x	o	x	x	o	o	x	x	x	x	x	o	x
# factors: 7	o	o	x	o	x	x	o	o	x	x	o	x	x	o	x
# factors: 8	o	o	x	o	x	x	o	o	x	x	o	x	o	o	x

This design is not recommended for more than 8 factors. See Reference (3).

There are many linear graphs for the L16. They are shown on the following pages along with the Interaction Table.

NOTE: o Means OK to use column
x Means do not use, involved with interactions from other columns.

L16 "Western Equivalent".

Column number:	1	2	3	4	5	6	7	8	9	10	11	12	13	14	15
Run Number	D	C	-CD	B	-BD	-BC	BCD	A	-AD	-AC	ACD	-AB	ABD	ABC	-ABCD
1	-	-	-	-	-	-	-	-	-	-	-	-	-	-	-
2	-	-	-	-	-	-	-	+	+	+	+	+	+	+	+
3	-	-	-	+	+	+	+	-	-	-	-	+	+	+	+
4	-	-	-	+	+	+	+	+	+	+	+	-	-	-	-
5	-	+	+	-	-	+	+	-	-	+	+	-	-	+	+
6	-	+	+	-	-	+	+	+	+	-	-	+	+	-	-
7	-	+	+	+	+	-	-	-	-	+	+	+	+	-	-
8	-	+	+	+	+	-	-	+	+	-	-	-	-	+	+
9	+	-	+	-	+	-	+	-	+	-	+	-	+	-	+
10	+	-	+	-	+	-	+	+	-	+	-	+	-	+	-
11	+	-	+	+	-	+	-	-	+	-	+	+	-	+	-
12	+	-	+	+	-	+	-	+	-	+	-	-	+	-	+
13	+	+	-	-	+	+	-	-	+	+	-	-	+	+	-
14	+	+	-	-	+	+	-	+	-	-	+	+	-	-	+
15	+	+	-	+	-	-	+	-	+	+	-	+	-	-	+
16	+	+	-	+	-	-	+	+	-	-	+	-	+	+	-

L16 (2^{15})

STRATEGIES for use: number of factors	Place Factor:	A	B	C	D	E	F	G	H	DEFINING CONTRAST
4	In Column:	8	4	2	1					This is full factorial
5	In Column:	8	4	2	1	15				1=ABCDE
6	In Column:	8	4	2	1	14	7			1=ABCE,BCDF,ADEF
7	In Column:	8	4	2	1	14	7	11		1=ABCE,BCDF,ACDG, ADEF,BDEG,ABFG,CEFG
8	In Column:	8	4	2	1	7	11	14	13	1=BCDE,ACDF,ABCG,ABDH, ABEF,ADEG,ACEH,BDFG, BCFH,CDGH,CEFG,DEFH, BEGH,AFGH,ABCDEFGH

Interaction Table for L16.

column:	2	3	4	5	6	7	8	9	10	11	12	13	14	15
1	3	2	5	4	7	6	9	8	11	10	13	12	15	14
2		1	6	7	4	5	10	11	8	9	14	15	12	13
3			7	6	5	4	11	10	9	8	15	14	13	12
4				1	2	3	12	13	14	15	8	9	10	11
5					3	2	13	12	15	14	9	8	11	10
6						1	14	15	12	13	10	11	8	9
7							15	14	13	12	11	10	9	8
8								1	2	3	4	5	6	7
9									3	2	5	4	7	6
10										1	6	7	4	5
11											7	6	5	4
12												1	2	3
13													3	2
14														1

Linear Graphs of Table L16.

(1)

a

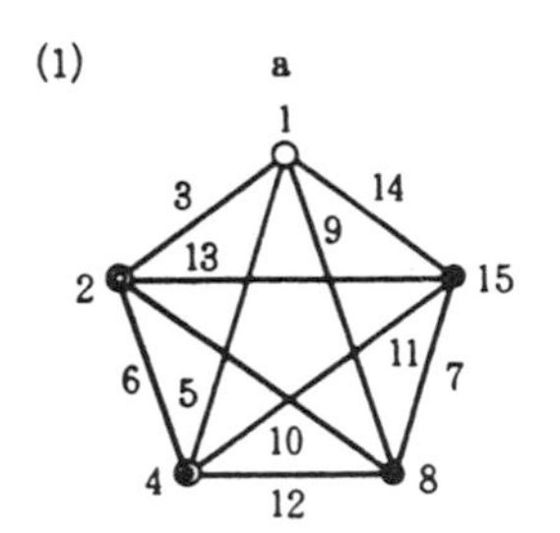

b

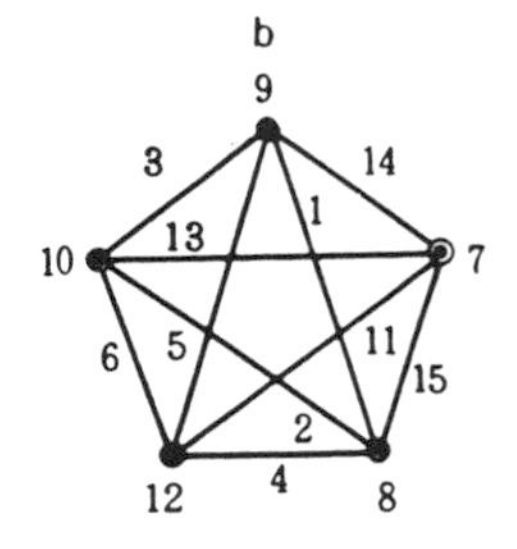

c

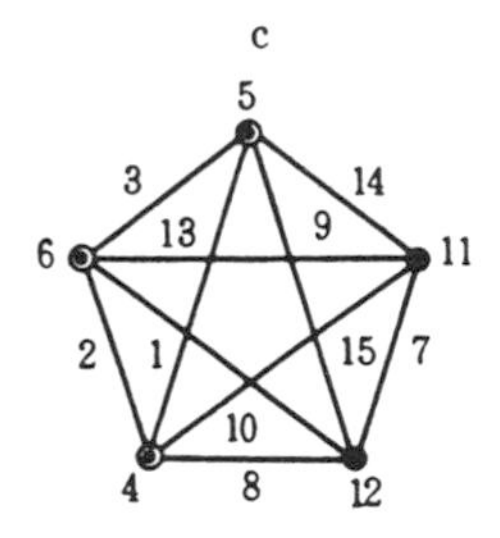

(2)

a

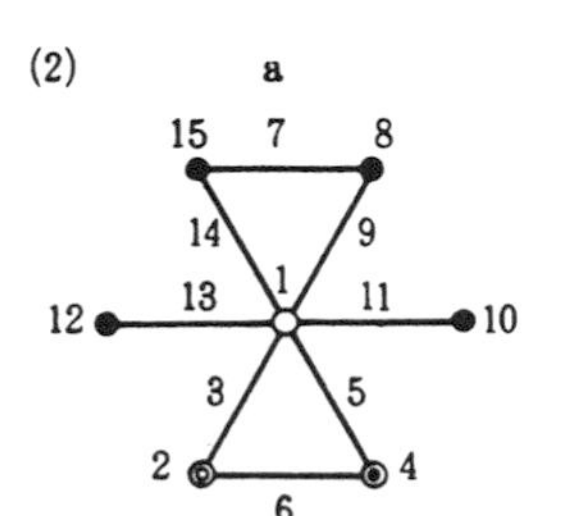

b

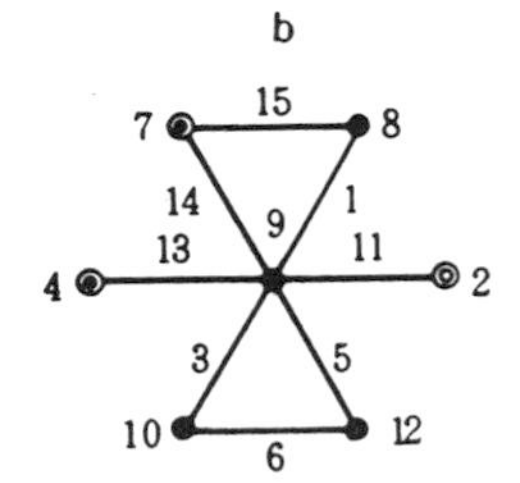

c

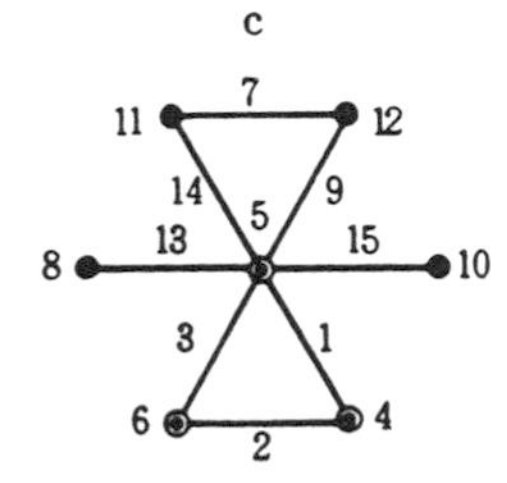

(3)

a

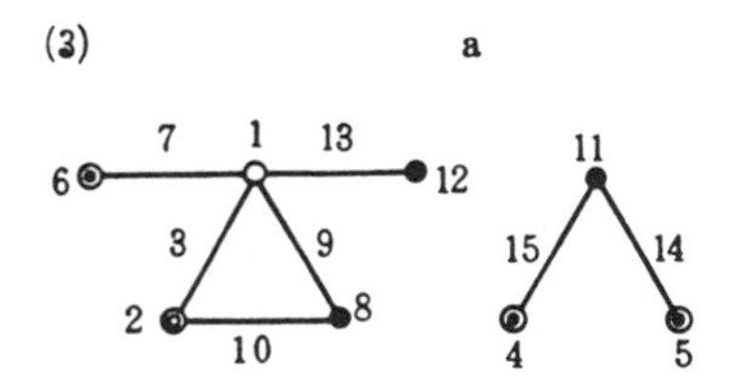

b

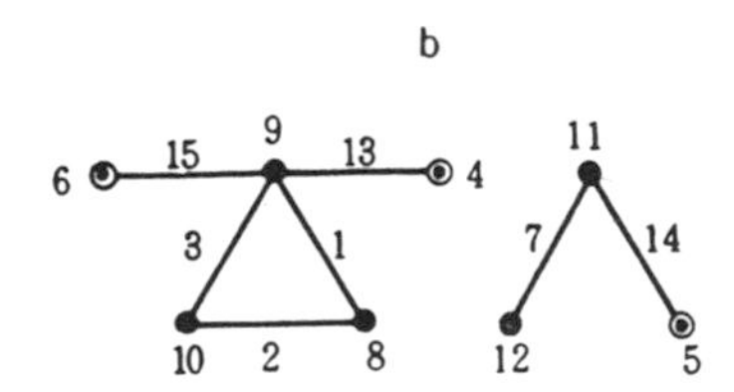

c

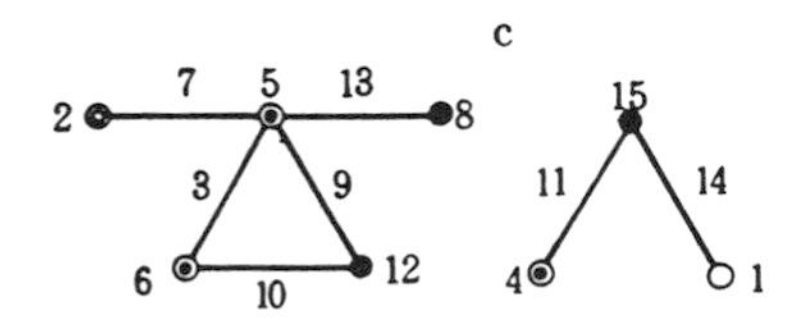

Linear Graphics of Table L16 (Continued).

(4) a

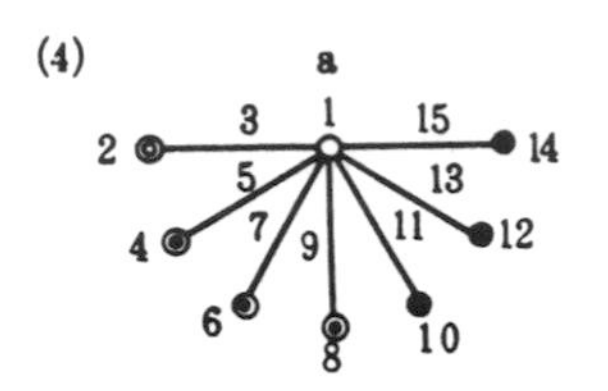

b

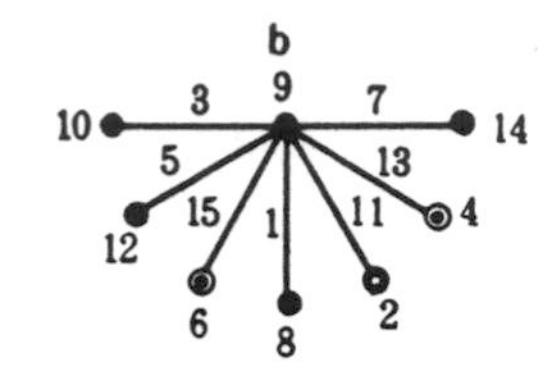

c

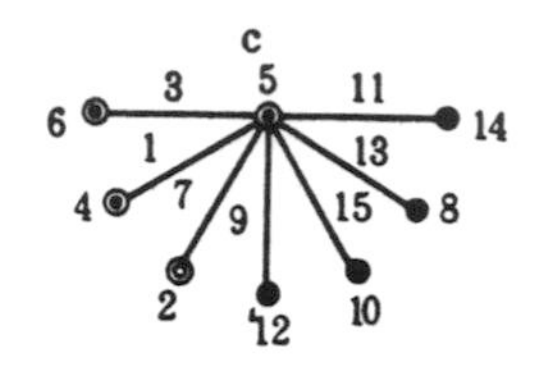

(5) a

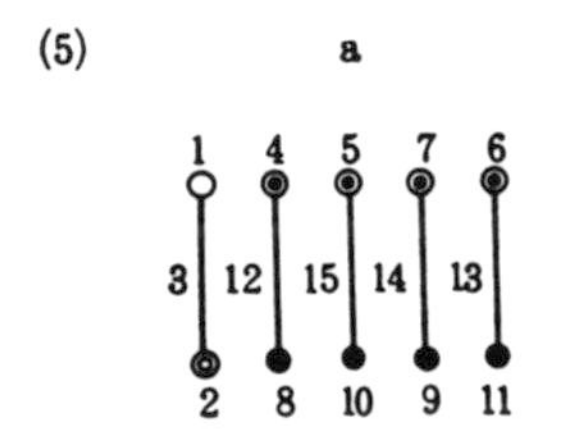

b

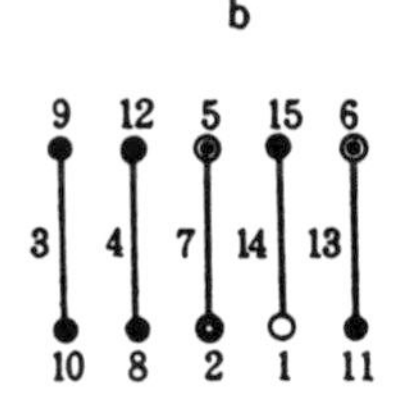

c

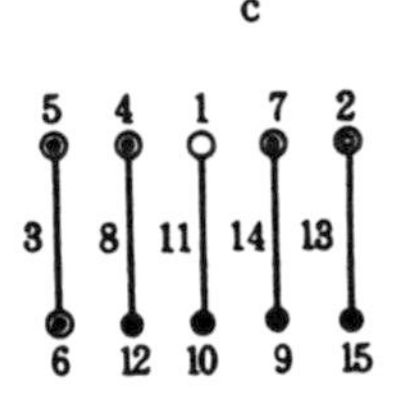

(6) a

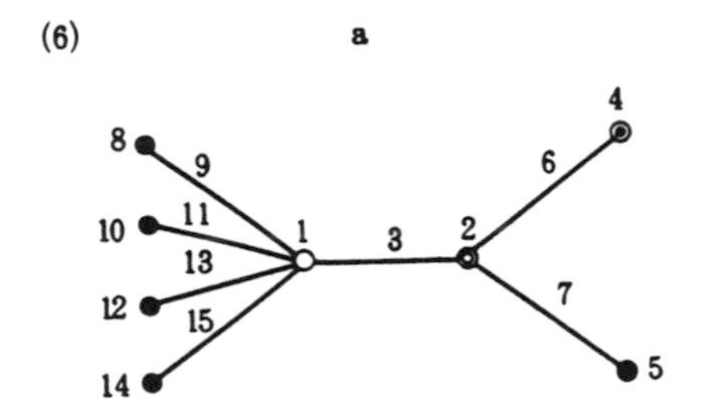

b

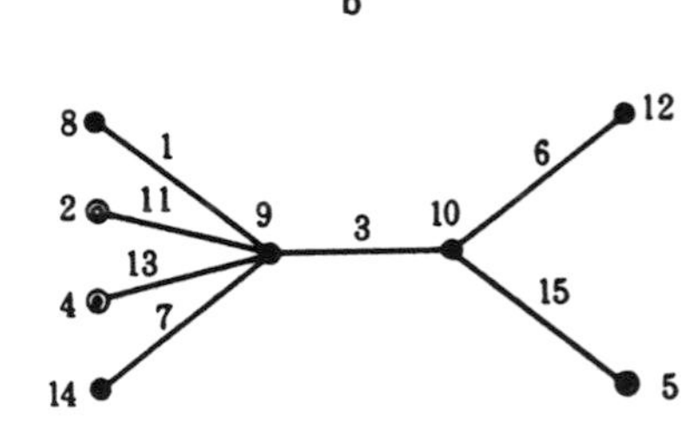

c

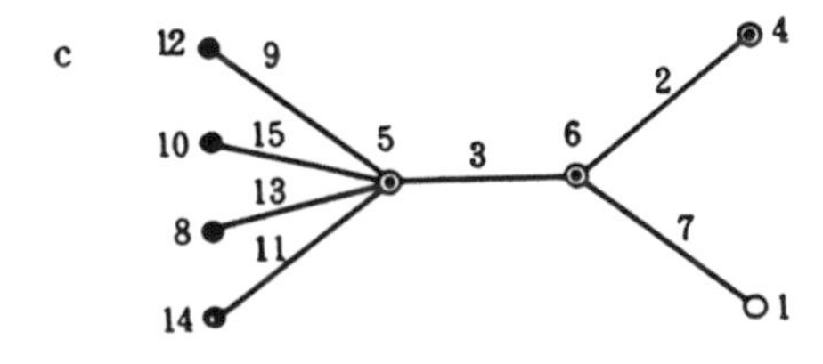

REFERENCES

1. American Society for Quality Control Annual Congress; May, 1989.
2. Box, G. E. P., Bisgaard, S., Fung, C. An Explanation and Critique of Taguchi's Contributions to Quality Engineering. Report Number 28, Center for Quality and Productivity Improvement, University of Wisconsin at Madison.
3. Barker, T. B. *Quality by Experimental Design.* Marcel Dekker, New York, 1985.

Index